Fuels and Petrochemicals Division

Held at the 2023 AIChE Annual Meeting

Orlando, Florida, USA
5-10 November 2023

ISBN: 978-1-7138-9306-6

Printed from e-media with permission by:

Curran Associates, Inc.
57 Morehouse Lane
Red Hook, NY 12571

Some format issues inherent in the e-media version may also appear in this print version.

Printed with permission by Curran Associates, Inc. (2024)

AIChE
120 Wall Street, FL 23
New York, NY 10005-4020

Phone: (800) 242-4363
Fax: (203) 775-5177

www.aiche.org

Additional copies of this publication are available from:

Curran Associates, Inc.
57 Morehouse Lane
Red Hook, NY 12571 USA
Phone: 845-758-0400
Fax: 845-758-2633
Email: curran@proceedings.com
Web: www.proceedings.com

TABLE OF CONTENTS

ADVANCES IN BIOFUELS PRODUCTION AND ALTERNATIVE FUELS I

111a Scaling Up Sustainable Petrochemicals 1
Hsiang Yee Hoekstra

111b Advancements in Biochemical Conversion of Different Lignocellulosic Feedstocks for Bioethanol Production: A Pilot Scale Approach 2
Narendra Naik Deshavath, Vijay Singh

111c Novel Ni-Siral Catalysts for the Production of Alkenes and Cyclo-Alkanes Via Ethylene Oligomerization 5
Kristinoba Olotu, Fernando Resende

111d Towards Developing a Pathway for Efficient Conversion of Syngas-Derived Olefins to Sustainable Aviation Fuel in High Yield and Selectivity 6
Anthony Giduthuri, Udishnu Sanyal, Mond F. Guo, Birgitte Ahring, Karthikeyan Ramasamy

111e Hydrothermal Liquefaction of the Brown Macroalgae Saccharina Latissima for the Production of Biofuel 7
Mayokun Oke, Fernando Resende

111f Separation of Torrefaction Bio-Oils Using Ammonia 9
Samuel Drummond, Thossaporn Onsree, Alexis Allegro, Jochen Lauterbach

111g Bio-Jet and Bio-Kerosene Production from Ozonolysis of High Oleic Oil 10
Junli Liu, Nathan S. Mosier, Elena Robles Molina, Xueli Chen, Gozdem Kilaz, Louis Edwards Caceres Martinez

111h Electrocatalytic Conversion of Phenol to Cyclohexane 11
Jeffrey Page, Amol Pophali, Tae Jin Kim, Stoyan Bliznakov, Julia A. Valla

111i Assessing the Long-Term Stability and Degradation of a Marine Advanced Biofuel Blend 13
Patritsia M. Stathatou

ADVANCES IN PETROLEUM PRODUCTION AND REFINING

252a Introduction to the Future of Petroleum Industry 14
M. R. Riazi, Vladimir Mahalec

252b Measurement of Calcite Solubility at Temperatures to 573K and Pressures to 68 Mpa 15
Hseen O. Baled, James Fazio, Ward A. Burgess, Isaac Gamwo

252c Providing Guidelines for Thermal Removal of Paraffin Deposits from Production Lines 16
Elijah Bell, Gabriel Santos, Nagu Daraboina, Cem Sarica

252d Automated in-Line Optical Trace Level Detection of Expanded Palette of Advanced Tracers 17
Sehoon Chang, Gawain Thomas, Wei Wang

252e Nanotags for Improved Cutting Depth Correlation 18
S. Sherry Zhu, Marta Antoniv, Gawain Thomas, Martin Poitzsch, Hyung Kwak

252f A Rapid Microwave-Assisted Bottle Test Method to Determine the Crude Emulsion Stability 20
Prasad Pawar, Clayton Jeffryes

252g Integration of Matrix Encoding in Transformer Model for Reaction Pathway Prediction in Oil Refining Process 21
Qiong Pan, Jie Li

252h Hydroxybetaine Zwitterionic Surfactants for Enhanced Oil Recovery: Compatibility with Acrylamide-based Copolymers and Rheological Properties 23
Muhammad Shahzad Kamal, Afeez Gbadamosi, Shirish Patil, Syed Hussain

252i Interfacial Behavior of Anionic Surfactant, Biopolymer and Hydrophilic Nanoparticles with Heavy Crude Oil for Enhanced Oil Recovery Applications in Sandstone Reservoirs 24
Rahul Saha

DEVELOPMENTS IN ALTERNATIVE FUELS AND ENABLING TECHNOLOGIES

343a Decarbonization of Long-Haul Trucking: LOHC Powertrains 26
Sayandeep Biswas, Kariana Moreno-Sader, Rob Jones, William Green

343b Solid-State Hydrolysis of Sodium Borohydride and Sodium Metaborate Hydrate 28
Geo Jong Kim, Savannah G. Hunt, Allison Millspaugh, Byung Uk Kim, Sang Bum Kim, Hyun-Tae Hwang

343c End-Use of Hydrogen in Decarbonizing Buildings and Industrial Manufacturing 29
Praveen Cheekatamarla

343d Hydroprocessing of Biomass-Derived Thermal Deoxygenation Oil for Producing Renewable Jet Fuel 30
Sampath Karunarathne, Sampath Gunukula, Hemant P. Pendse, Clayton Wheeler

343e Advancing the State of Hydrogen Systems Quantitative Risk Assessment: Design and Requirements of a Hydrogen Component Reliability Database (HyCReD) 31
Ahmad Al-Douri, Katrina Groth, Kevin Hartmann, Genevieve Saur, William Buttner

343f Ring-Opening of Thermal Deoxygenation (TDO) Oil over Supported Metal Catalysts to Improve Diesel Engine Properties 34
Oluwaseun Salami, Matthew Kline, Sampath Karunarathne, Clayton Wheeler

343g NOx-to-Ammonia for Circular Ammonia-Fueled Container Ships 35
Margarita Charalambous, Alexander Schoch, Gonzalo Guillén-Gosálbez

343h Aspen Plus Simulation for Syngas Conversion with High Pressure Plasma 38
Dylan Broussard, Varshith Vijayan, Ramalingam Subramaniam

343i Gaseous Fuels Derived from Landfill-bound Waste- MSW and Plastics- for a Stationary Power Generation Engine: A Performance and Emissions Analysis 39
Lauren Creadore, Marco J. Castaldi

343j Influence of CO_2 Concentrations on Chlorella Kessleri Cultivation, Growth, and CO_2 Fixation 43
Shaikh Razzak

DEVELOPMENTS IN ELECTROCHEMICAL REACTORS, FUEL CELLS, AND ELECTROLYZERS I

369b Effect of Gas Composition on Co-Electrolysis Performance of Solid Oxide Electrolyzer Cells............... 44
Masashi Kishimoto, Riki Nakamura, Hiroshi Iwai

369c Advances in High Temperature Electrolysis for Sustainable Hydrogen Production: Insights from Idaho National Laboratory's Renewable Energy Research Portfolio.. 47
Joshua Gomez

369d Development of a New Packed-Bed Type Electrolyzer for the Production of CO or Syn-Gas through Electrochemical CO_2 Reduction Reaction.. 48
Donggu Han, Ung Lee

369e Assessing and Improving Electrode Structures for Scalable CO_2 Reduction Using in Operando electrochemical Techniques.. 49
Tim Van Cleve, Prantik Saha, Danielle Henckel, Leiming Hu, K. C. Neyerlin

DEVELOPMENTS IN SHALE GAS AND NATURAL GAS

449a Introduction to Natural Gas Production and Processing.. 50
M. R. Riazi

449b Gas Hydrate Based Solidified Natural Gas (SNG) Technology Incorporating Dual-Function Promoters for Enhanced Commercialization Potential.. 51
Gaurav Bhattacharjee, Ye Zhang, Praveen Linga

449c Pore-Scale Methane Hydrate Depressurization in Confined Pores for Efficient Methane Recovery.. 52
Qian Ouyang, Jyoti Shanker Pandey, Nicolas Von Solms

449d Determination of Local Crystal Orientation of Methane Hydrates in Tailored Confined Media Using Polarized Raman Spectroscopy... 53
Mrityunjay Sharma, Wiwin Febrianto, Yuan Yang, Carolyn Koh, Ryan Hartman

449e Commercial Applications of Petroleum Data Analytics Using Shale Oil and Gas Data Sets from Permian Operator.. 54
Vyacheslav Romanov, Guoxiang Liu, Shahab D. Mohaghegh, Xiongjun Wu, Derek M. Vikara, Kolawole Bello

449f Tuning Activity and Selectivity in the Partial Oxidation of Methane over Iridium and Ruthenium Mixed Oxides Supported on Titanium Dioxide Nanoparticles 55
Helena Hagelin Weaver, Li-Yin Hsiao

RECENT DEVELOPMENTS IN FUEL PROCESSING FOR HYDROGEN PRODUCTION I

551a Heavy Oil Reforming in a Dual Circulating Fluidized Bed Reactor 56
Girish Srinivas, Steve Schwab

551b Fast Start-up, Fuel Flexible Cpox Reformer for Efficient Syngas Generation from Sustainable Feedstocks 59
Hani Hawa, Subir Roychoudhury, Saurabh A. Vilekar, Christian Junaedi, John Hong

551c Maximizing Conversion and Coking Stability for Hydrogen Production Via Dry Reforming of Methane over a Ni/Zeolite Catalyst 60
Janet Baffoe, Xuan Pham, Corey Leclerc

551d High Conversion of Two Major Greenhouse Gases to Syngas through Dry Reforming Reaction over Efficient Nickel-Hydroxyapatite Catalyst 61
Hanaa Hassini, Esma Ines Achouri

551e Dry Reforming of Methane over Ni Catalysts on High Oxygen Ion Conductivity Supports 62
Xuan Pham, Janet Baffoe, Corey Leclerc

551f Syngas Production from Methane Steam Reforming over Nickel Supported Hydroxyapatite Catalyst 63
Sabrina Bahia Karakache, Esma Ines Achouri, Nicolas Abatzoglou

551g Crude Methanol from Alternative Energy Source As Low-Cost, Low-Carbon Hydrogen Carrier 64
Hsiang-Sheng Chen, Kritika Sharma, Vijay Ramani, Nazim Muradov, Paul Yelvington, Edwin Yik

551h Energy-Saving Electrochemical Hydrogen Production Coupled with Value-Added Isobutanol Oxidation to Isobutyric Acid 66
Ruiqi Du, Siqi Zhao, Kaizheng Zhang, Shuairen Qian, Yi Cheng

INTEGRATED BIOREFINERIES: TECHNOLOGIES AND TEA/LCA I

507a TEA and LCA of Renewable Natural Gas: Upgrading Biogas from Anaerobic Co-Digestion of Prairie Grass with Hydroxycinnamic Co-Products 67
Katherine Wild, Mark Mba Wright, Lisa Schulte

507b Low-Dimensional Techno-Economic Cost Model for Expedited Hydrothermal Liquefaction Process Evaluation 68
Muntasir Shahabuddin, Eduardo Italiani, Andrew R Teixeira, Nikolaos Kazantzis, Michael T. Timko

507c Catalytic Hydrodeoxygenation of Hydrothermal Liquefaction (HTL) Derived Bio-Oil from Corn Stover for Sustainable Aviation Fuels 70
Zoe Fickbohm, Bharathkiran Maddipudi, Rajesh Shende, Khang Huynh

507d TEA/LCA of Integrated Processing of Corn Stover for Fuels and High Value Products 71
Bharathkiran Maddipudi, Khang Huynh, Vinod Amar, Katelyn Shell, Anuj Thakkar, Anuradha Shende, John E. Aston, Sandeep Kumar, Ram Gupta, Rajesh Shende

507e Integration of Biomass Conversion and Solar Hydrogen Generation Technologies to Produce Renewable Fuels 72
Rahul Bhosale, Sayma Akhter, Gopalakrishnan Kumar

POSTER SESSION: FUELS AND PETROCHEMICALS DIVISION

83a Evolving Robust and Interpretable Enzymes for the Bioethanol Industry 84
Anni Li, Yijie Sheng, Minghui Wang, Xiujuan Li, He Huang

83b In-house Developed Resilient and Sustainable Material for Enhanced Oil Recovery 87
Syed Muhammad Shakil Hussain, Muhammad Shahzad Kamal, Afeez Gbadamosi, Shirish Patil, Dhafer Al-Shehri

83c Assessing the Technical Feasibility of Steam Hydrogasification of Agriculture-Derived Wastes for Renewable Fuel Production in the Leading Agriculture Region 88
Zhongzhe Liu, Chan Seung Park, Partho Roy, Xin Fan, Marco Ceja, Rosely Ayala, Robert Lozano, Serina Ishida

83d Multiphase Particle in Cell Simulation Study of Ammonia Co-Combustion in a Circulating Fluidized Bed Combustor 89
Jester Lih Jie Ling, Han Saem Park, Ha Eun Lee, Seung Seok Oh, Hyun Jun Park, Sang Mun Jeong, See Hoon Lee

83e Hydrogen Production By Oxidation of Aluminum Nanopowder in Water Under the Action of Laser Pulses 90
Yaroslav Kraft, Boris Aduev, Denis Nurmukhametov, Gennadiy Belokurov, Natalia Nelubina, Zinfer Ismagilov

83f Performance Targets for Oxidative Coupling of Methane from Techno-Economic Profiling 95
Faisal Ashour, Mamoun Al-Rawashdeh, Patrick Linke

PROPERTIES AND PHASE EQUILIBRIA FOR FUELS AND PETROCHEMICALS

529a Creating a Ternary Phase Diagram for Pyrene and Its Oligomers: The Effect of Oligomeric Composition on Liquid Crystalline Phase Behavior 96
Jefferson Giles II, Mark C. Thies

529c Development of a Robust Thermodynamic Model to Predict the Onset Crystallization Temperature of Fatty Acid Methyl Esters(FAMEs) 97
Junli Liu, Bernard Tao

529d Predictive Parametrization of PC-SAFT EoS By Neural Network Ensembles Based on Molecular Fingerprints 98
Jonas Habicht, Christoph Brandenbusch, Gabriele Sadowski

529e Heat Capacity of Saturated and Compressed Liquid Dimethyl Ether at Temperatures from (132 to 345) K and at Pressures to 35 Mpa: Measurements and Derived Equation of State 99
Joseph Magee, Jiangtao Wu

UNCONVENTIONAL OIL AND NATURAL GAS: SCIENCE AND TECHNOLOGY ADVANCEMENT

582a Competitive Adsorption and Displacement Behavior of CO_2/CH_4 Binary Mixture in Shale Matrix with Heterogeneous Surfaces 100
Jingyue Sun, Cong Chen, Zhehui Jin, Weizhong Li

582c Integrating Public and Private Data for Modeling and Optimization of Shale Oil and Gas Production 102
Vyacheslav Romanov, Derek M. Vikara, Kolawole Bello, Shahab D. Mohaghegh, Guoxiang Liu, Luciane Cunha

VALUE-ADDED CHEMICALS FROM NATURAL GAS

591a Direct Conversion of Methane on Metal-Oxide and Oxide-Oxide Interfaces: Fundamental Studies on the Production of Hydrogen and Methanol 103
Jose Rodriguez

591b Rich CH_4 Oxidation on $Pt/Pd/Al_2O_3$ Monolith: Performance, Multiplicity, and Spatial Effects 104
Jonathan Ratcliff, Michael Harold

591c Aiding Catalysis with Additive Manufacturing: Fabrication and Optimization of Hierarchical Membrane/Catalyst Systems for Oxidative Coupling of Methane 106
James Wortman, Valentina Omoze Igenegbai, Ali Hussain Motagamwala, Rawan Almallahi, Suljo Linic

591d Chemical Looping for the Production of Value-Added Oxygenates from CO_2 and Natural Gas 107
Sherafghan Iftikhar, Luke Neal, Junchen Liu, Kyle Vogt-Lowell, Vasudev Haribal, Andrew Tong, Fanxing Li

591e Supported Catalytically Active Liquid Metals Solutions (SCALMS) As Catalysts for Ethane Dehydrogenation 108
Aime Laurent Twizerimana, Gia Monte, Aaditya Hari Bharanidharan, Kaylen Ocampo, Kyle McGaughy, Mohammad S. Masnadi, Goetz Veser

591f Oxidative Ethane Dehydrogenation Under Thermal and Microwave Heating over Ga/ZSM-5, Gapt/ZSM-5, and Microwave Synthesized Ga-ZSM-5 110
Ashley Caiola, Brandon Robinson, Sean Brown, Xiaoyan Wang, Yuxin Wang, Jianli Hu

591g Portable Process Development for Methanol Synthesis from Stranded Methane 111
Hsiang-Sheng Chen, Nazim Muradov, Wendy Zhou, Paul Yelvington, Edwin Yik

591h Development of a Zero-Carbon Methane Dehydroaromatization Process Design 113
Raid J. Hassiba, Mamoun Al-Rawashdeh, Patrick Linke

ADVANCES IN BIOFUELS PRODUCTION AND ALTERNATIVE FUELS II

200d Renewable Blendstocks from Hazardous Industrial Wastes Via Hydrothermal Liquefaction 114
Harisankar S, R. Vinu

200b Mechanical Preprocessing of Non-Recyclable Municipal Solid Wastes Towards Advanced Biorefinery Feedstock 116
Nepu Saha, Jordan Klinger, Wencheng Jin, Neal Yancey, Md Tahmid Islam, Toufiq Reza, Abdallah Ikbarieh, Sheng Dai, Megan Herrera

200c Process Simulation and Yield Optimization of Bio-Oil from Hydrothermal Liquefaction of Macroalgae(Kelp) Using Mixtures of Model Compounds. 117
Micheal Asama, Fernando Resende, Aaditya Khanal

200a Renewable Liquefied Natural Gas As Marine Fuel: Process Simulation-Based Analysis of Different Production Routes Including Energetic, Environmental and Economic Aspects 118
Elena Barbera, Andrea Mio, Michele Adinolfi, Maurizio Fermeglia, Alberto Bertucco

200f Catalytic Fast Hydropyrolysis of Beetle Killed Trees- Pine (pinus taeda), Ash(fraxinus) and Redbay (persea borbonia) for the Production of Transportation Biofuel 119
Oluwanisola Makinde, Fernando Resende

200g Engineering the Kinetics of Acid Hydrolysis of Corn COB to Glucose for BIO-Ethanol Production 120
Okechukwu Onyelucheya, Jude C Obijiaku, Chioma M Onyelucheya

DEVELOPMENTS IN ELECTROCHEMICAL REACTORS, FUEL CELLS, AND ELECTROLYZERS II

417a Determination of the Cathode Degradation Mechanism in Hydroxide Exchange Membrane Fuel Cells 121
Catherine M. Weiss, Brian P. Setzler, Yushan Yan

417b Dynamic Modeling and Control of a PEM Electrolyzer for Solar Photovoltaic Power Smoothing 122
Yifan Wang, James Fenton, Paul Brooker

417c Phosphonic Acid Ionomers for Use As Free Radical Scavengers in Proton Exchange Membrane Fuel Cells 123
Alisa Chakraborty, Christopher Arges

417d Enhanced Power Management for Direct Methanol Fuel Cell/Battery Hybrid System By Improved Cascade Control Structure 124
Zuhair Alyousef, Majed Aljeshi

417e Analysis of Mixed Potential at the Cathode of a Direct Methanol-Hydrogen Peroxide Fuel Cell for Enhanced Performance 128
Kritika Sharma, Vijay Ramani

RECENT DEVELOPMENTS IN FUEL PROCESSING FOR HYDROGEN PRODUCTION II

570a Carbon Catalyzed Decarbonization of Natural Gas for Clean Hydrogen and Solid Carbon 129
Randall Vander Wal, Mpila Nkiawete

570b Process Intensified Modular Production of Clean Hydrogen and Carbon Nanomaterials from Natural Gas By Microwave Catalytic Pyrolysis 132
Changle Jiang, Jianli Hu, Sonit Balyan, Brandon Robinson, Juan A. Lopez-Ruiz, Yuan Jiang, Robert A. Dagle, Sherafghan Iftikhar, Luke Neal, Fanxing Li, Wenyuan Li, Xingbo Liu, George Skoptsov

570c Tuning Particle ALD Catalysts for Methane Pyrolysis 134
Jessica Hauck, Kent J. Warren, Boning Wang, Samantha Harshberger, Mija H. Hubler, Andrew Broerman, Theodore Champ, Christopher Gump, Alan Weimer

570d Tailoring the Combustion Synthesis of Fe-Al-O Catalysts for Microwave-Assisted Production of Hydrogen from Fossil Fuels 135
Zachary Chanoi, Laura A. Martinez Espinoza, Victoria Reyes, Evgeny Shafirovich

570e Microwave-Assisted Moving-Bed Gasifier Design for Solid Fuel Conversion 136
Xinwei Bai, Pranjali Muley

570f Modularization Chemical Looping System for Hydrogen Production from Carbonaceous Solid Feedstock Along with CO2 Capture and Additional Heat Recovery 137
Rushikesh Joshi, Eric Falascino, Andrew Tong, Liang-Shih Fan

570g Methane Looping Reforming to Syngas: Optimal Process Conditions and Carrier Lifetime Assessment 139
Alba Storione, Matteo Minelli, Mattia Boscherini, Elena Landi, Francesco Miccio, Ferruccio Doghieri

570h Sustainability Analysis of Hydrogen Production Processes for Maritime Transportation....................... 141
Andrea Mio, Elena Barbera, Alessandro Massi Pavan, Alberto Bertucco, Maurizio Fermeglia

INTEGRATED BIOREFINERIES: TECHNOLOGIES AND TEA/LCA II

526b Preliminary Economic Assessment of Microalgae Via Hydrothermal Liquefaction with Co-Production of Protein Concentrates 142
Yiling Xu, Yunhua Zhu, Peter Valdez, Lesley J. Snowden-Swan

526c Sustainable Biolubricant Production from Lignin and Waste Cooking Oil 143
Kyoko Hirayama, Elena Barbera, Randy Maglinao, Alberto Bertucco, Sandeep Kumar

526d A Process for Co-Conversion of Bio-Oil in Delayed Coker Unit As Viable Option for Achieving Net Zero Goals.. 144
Satyen Kumar Das, Shivam Ashok Dixit, Pradeep P. R., Sarvesh Kumar, Madhusudan Sau, S. S. V. Ramakumar

IN THE SPIRIT OF INNOVATION

Evolution and Development of Fluidized Catalytic Dehydrogenation (FCDh) Technology 145
Matt Pretz

483b Ethylene Oxide Catalyst and Process Innovations: A Successful Journey Towards Asustainable Industrial Process 146
Madan M. Bhasin

483c Imagine This a Tribute to Bipin Vora on His 80th Birthday 147
James Johnson

483d Strategies for Improved Thermal Stability and Facile Regeneration of Alkane Dehydrogenation Catalysts.. 148
Abhaya Datye, Ryan Alcala, David P. Dean

483e Methanol-to-Olefins (MTO). Bipin Vora's Key Role in the Development of the UOP/Hydro MTO Process Technology 149
Steinar Kvisle, James Johnson

Author Index

111a Scaling Up Sustainable Petrochemicals

Hsiang Yee Hoekstra

One of the main aspects of promising solutions for reducing greenhouse gas emissions and promoting sustainable energy production has been green petrochemicals. Derived from renewable and sustainable sources, such as biomass, plastics and civic waste streams, petrochemicals are processed using advanced technologies to produce fuels and chemicals with low environmental impact. To replace conventional fossil fuels from the transportation and industrial sectors, renewable gasoline, diesel, and sustainable aviation fuels (SAFs) have been proposed as the candidates with the highest potential.

At SwRI, we have integrated teams working on all facets of renewable petrochemicals and related infrastructure developed for real-world fuel producers. Our expertise ranges over thermochemical methods for conventional and unconventional feedstock cracking and conversions facilitated by scaled-up isomerization and hydrotreating technologies to obtain refined fuels while achieving the desired properties and quality standards. Simultaneously, SwRI also works on engine testing and required adaptations for such renewable fuels. Renewable gasoline and diesel are characteristically like their fossil fuel counterparts and thus can be used as standalone fuels or blended with fossil fuels in internal combustion engines.

Sustainable aviation fuels require a more stringent approach to design and performance requirements of the aviation industry. We have prior experience in producing SAFs from feedstocks like biomass, municipal solid waste and plastics as well as upgrading from petroleum sources. SAFs have been approved for commercial flights and have already demonstrated significant reduction in greenhouse emissions. Through numerous and diverse projects completed over decades SwRI promotes an achievable and sustainable energy and material economy. This presentation will cover the different facets of scaling-up and upgrading different feedstocks to renewable petrochemicals while competing with fossil-based petrochemicals.

111b Advancements in Biochemical Conversion of Different Lignocellulosic Feedstocks for Bioethanol Production: A Pilot Scale Approach

Narendra Naik Deshavath, Vijay Singh

The conversion of lignocellulosic biomass into bioethanol and its blending with gasoline for the transportation sector can reduce the dependency on fossil fuels. However, several factors are affecting the commercialization of lignocellulosic biorefinery, including the enzyme costs, incomplete conversion of polymeric carbohydrates into monomeric sugars, and the formation of fermentative inhibitors. Fermentative inhibitors deter the fermentation efficiency of hydrolysates which leads to reduced yield and productivity [1]. Generally, fermentative inhibitors can be classified into three categories: process-derived, inherent, and supplemented. Process-derived fermentative inhibitors include Furfural, 5-Hydroxymethylfurfural (5-HMF), levulinic acid, and formic acid. These are formed during the pretreatment of lignocellulosic biomass under acidic conditions with moderate to high temperatures [2]. Furfural is a decomposed product of xylose and arabinose, and 5-HMF is a decomposed product of glucose and fructose. Further decomposition of 5-HMF forms levulinic acid and formic acid, whereas furfural forms only formic acid. Apart from these, acetic acid, glucuronic acid, ferulic acid, vanillic acid, coumaric acid, benzoic acid, hydroxybenzoic acid, and syringaldehyde are the inherent inhibitors derived during the hydrolysis of lignocellulosic biomass [3–5]. Acetic acid, glucuronic acid, and ferulic acid are the structural constituents of hemicellulose, which is linked with the xylan backbone [6]. Moreover, vanillic acid, coumaric acid, syringaldehyde, etc., are the structural components of lignin [7]. Finally, citrate buffer and sulfate ions are the unavoidable major supplement inhibitors used during the enzymatic hydrolysis of pretreated lignocellulosic biomass. Citrate buffer plays a vital role during the enzymatic hydrolysis biomass, which restricts the pH change of the hydrolysis medium [8]. Generally, 50 mM citrate buffer strength is used to maintain the enzymatic hydrolysis medium pH between 4.8 and 5.5; it deters microbial metabolic growth in the subsequent fermentation of hydrolysates [9, 10]. Therefore, in addition to the process-derived and inherent inhibitors, 50 mM citrate buffer strength also acts as an inhibitor during hydrolysate fermentation.

Several pretreatment technologies were investigated to deconstruct the complex network of lignocellulosic biomass. Dilute sulfuric acid and alkali are the most widely used inorganic catalytic agents for the pretreatment of lignocellulosic biomass. Dilute sulfuric acid pretreatment hydrolyzes most of the hemicellulose fraction but forms sugar decomposition products [11, 12]. In contrast, alkaline pretreatment delignifies the lignocellulosic biomass, which enhances enzymatic digestibility but eliminates most of the hemicellulose fraction along with the minor fraction of cellulose, which eventually affects the total ethanol yield per dry ton of lignocellulosic biomass [13]. Moreover, washing pretreated biomass required considerable water to remove residual acid or base before the enzymatic hydrolysis [14]. Several studies also performed enzymatic hydrolysis of dilute sulfuric acid pretreated biomass by directly adjusting the pH at 4.8 to 5.5 with ammonium hydroxide instead of washing [15]. However, it forms sulfate ions that deter microbial metabolic growth, leading to low ethanol yield and productivity [16]. In addition, high levels of ammonium salts in the stillage required a specialized wastewater treatment section in the downstream processing [15].

Hydrothermal pretreatment is a leading approach for the possible commercialization of 2G biorefinery because they avoid the addition of chemical catalysts that add operating costs and complicate downstream processing. Hydrothermal pretreatment deconstructs the conglomerate structure of lignocellulosic biomass and makes accessibility to the enzymes to hydrolyze hemicellulose and cellulose into monomeric sugars (glucose, xylose, and arabinose), and subsequent microbial conversion of sugars into bioethanol [17]. In this regard, we developed a two-stage pretreatment process whereby biomass is treated at 190 °C for 10 min at 50% (w/w) moisture, followed by disc milling [18]. Disc milling further opens the cellulose fibers and increases sugar yields in the subsequent enzymatic hydrolysis. This process has since been scaled up with a continuous steam explosion reactor using bioenergy sorghum as the feedstock [19]. Fed-batch enzymatic hydrolysis of pretreated biomass at high solids loading (50% w/v) yields 230 g/L sugar concentration by the cellulose and hemicellulose hydrolysis efficiency of 65% and 85%, respectively [20]. Another significant advantage of the hydrothermal pretreatment followed by mechanical refining is that it minimizes process-derived fermentative inhibitors like furfural, 5-Hydroxymethylfurfural (5-HMF), levulinic acid, and formic acid. Our approach also reduced the usage of citrate buffer strength to 0.5 mM during the enzymatic hydrolysis process because the hydrothermal pretreatment was conducted at a lower severity without using acid catalysts and limited the formation of sugar decomposition products [21]. However, inherent inhibitors are still released from the lignin and hemicellulose fractions [3–6].

Hydrothermal pretreatment followed by a mechanical refining approach has been investigated on different lignocellulosic feedstocks such as oilcane 1566 (a genetically modified sugarcane that produces lipids along with sugars), energy cane, and miscanthus × giganteus. In this study, we made more advancements in the enzymatic hydrolysis step, which was carried out in distilled water without controlling the pH by acids or bases, and the sugar yields were compared with standard citrate buffer strength (50 mM). Fed-batch enzymatic hydrolysis was conducted with 50% (w/v) solid loading at 50 °C for 96 h using cellulase (NS 22257) and hemicellulase (NS 22254). In the 50 mM citrate buffer containing enzymatic hydrolysis medium, 215.28±0.69 g/L, 249.2±0.44 g/L, 238.52±0.56 g/L, and 192.78±0.99 g/L sugar concentrations were obtained from bioenergy sorghum, oilcane 1566, energy cane and miscanthus × giganteus, respectively. Whereas 215.05±0.62 g/L, 239.61±0.98 g/L, 231.85±1.45 g/L, and 184.65±0.44 g/L sugar concentrations were attained from bioenergy sorghum, oilcane 1566, energy cane and miscanthus × giganteus, respectively without using citrate buffer. Relatively similar sugar concentrations were obtained either using 50 mM citrate buffer or without citrate buffer. Further, enzymatic hydrolysates derived without citrate buffer were fermented by commercial genetically modified xylose-fermenting *Saccharomyces cerevisiae* for bioethanol production. Fermentation of bioenergy sorghum, oilcane 1566, energy cane, and miscanthus × giganteus hydrolysates produced 78.56±1.44 g/L, 59.23±0.66 g/L, 68.28±3.03 g/L, and 51.18 g/L bioethanol, respectively.

References

1 J. Baeyens, Q. Kang, L. Appels, R. Dewil, Y. Lv and T. Tan, *Prog. Energy Combust. Sci.*, 2015, **47**, 60–88.

2 N. N. Deshavath, N. V. Mogili, M. Dutta, L. Goswami, A. Kushwaha, V. D. Veeranki and V. V. Goud, in *Waste-to-Energy Approaches Towards Zero Waste*, Elsevier, 2022, pp. 57–85.

3 G. Wan, Q. Zhang, M. Li, Z. Jia, C. Guo, B. Luo, S. Wang and D. Min, *J. Agric. Food Chem.*, 2019, **67**, 10116–10125.

4 T. Pisithkul, T. B. Jacobson, T. J. O'Brien, D. M. Stevenson and D. Amador-Noguez, *Appl. Environ. Microbiol.*, 2015, **81**, 5761–5772.

5 L. Qin, W. C. Li, L. Liu, J. Q. Zhu, X. Li, B. Z. Li and Y. J. Yuan, *Biotechnol. Biofuels*, 2016, **9**, 1–10.

6 N. N. Deshavath, V. D. Veeranki and V. V. Goud, in *Sustainable Bioenergy*, eds. M. Rai and A.

Ingle, Elsevier, 1st edn., 2019, pp. 1–19.

7 B. A. McKinley, S. N. Olson, K. B. Ritter, D. W. Herb, S. D. Karlen, F. Lu, J. Ralph, W. L. Rooney and J. E. Mullet, *PLoS One*, 2018, **13**, 1–20.

8 A. Arora and D. J. Carrier, *ACS Sustain. Chem. Eng.*, 2015, **3**, 2423–2428.

9 N. N. Deshavath, V. V. Goud and V. D. Veeranki, *Fuel*, 2021, **287**, 119545.

10 N. N. Deshavath, V. V. Goud and V. D. Veeranki, *J. Environ. Chem. Eng.*, 2021, **9**, 105696.

11 N. N. Deshavath, M. Mohan, V. D. Veeranki, V. V. Goud, S. R. Pinnamaneni and T. Benarjee, *3 Biotech*, , DOI:10.1007/s13205-017-0752-3.

12 N. N. Deshavath, S. Mahanta, V. V. Goud, V. V. Dasu and S. R. P., *J. Environ. Chem. Eng.*, , DOI:10.1016/J.JECE.2018.08.002.

13 H. Chandel, P. Kumar, A. K. Chandel and M. L. Verma, *Biomass Convers. Biorefinery*, , DOI:10.1007/s13399-022-02746-0.

14 L. Tao, D. Schell, R. Davis, E. Tan, R. Elander and A. Bratis, *NREL 2012 Achievement of Ethanol Cost Targets : Biochemical Ethanol Fermentation via Dilute-Acid Pretreatment and Enzymatic Hydrolysis of Corn Stover*, 2014.

15 and A. A. D. Humbird, R. Davis, L. Tao, C. Kinchin D. Hsu, *Process Design and Economics for Biochemical Conversion of Lignocellulosic Biomass to Ethanol: Dilute-Acid Pretreatment and Enzymatic Hydrolysis of Corn Stover*, 2011.

16 E. Casey, N. S. Mosier, J. Adamec, Z. Stockdale, N. Ho and M. Sedlak, *Biotechnol. Biofuels*, 2013, **6**, 1–10.

17 N. Mosier, C. Wyman, B. Dale, R. Elander, Y. Y. Lee, M. Holtzapple and M. Ladisch, *Bioresour. Technol.*, 2005, **96**, 673–686.

18 S. M. Kim, B. S. Dien and V. Singh, *Biotechnol. Biofuels*, 2016, **9**, 1–15.

19 M. H. Cheng, B. S. Dien, D. K. Lee and V. Singh, *Bioresour. Technol.*, 2019, **289**, 121663.

20 M. H. Cheng, H. J. Kadhum, G. S. Murthy, B. S. Dien and V. Singh, *Bioresour. Technol.*, 2020, **318**, 124051.

21 N. N. Deshavath, B. S. Dien, P. J. Slininger, Y. Jin and V. Singh, *Fermentation*, 2023, **9**, 1–19.

111c Novel Ni-Siral Catalysts for the Production of Alkenes and Cyclo-Alkanes Via Ethylene Oligomerization

Kristinoba Olotu, Fernando Resende

The adverse environmental impact of fossil fuels has expedited the development of alternative energies specifically biofuels to gain traction. Constituents of jet fuel such as linear and cyclic hydrocarbons can be produced from sustainable sources which have the ability to reduce soot generation, prevent leaks in fuel systems and increase energy content. Research has shown that solid heterogenous catalysts play a pivotal role in reactions that promote yield and selectivity of linear and cyclic hydrocarbon compounds which are constituents of jet fuel. The oligomerization of ethylene is one of the main routes considered for the conversion of ethylene into these hydrocarbon mixtures. Commercial ethylene oligomerization process uses organometallic based homogeneous catalyst, in which case it is difficult and expensive to separate catalyst from products. Conversely, the development of heterogeneous catalyst with the ability to promote coke dissolution preventing pore plugging and consequent catalyst deactivation is a paramount factor in this reaction. Previous studies have reported production of 43% yield of liquid products with supercritical ethylene at 65 bar and 200 with 2.8wt % nickel loading of a novel Ni-SIRAL 40 HPV catalyst, consequent of its large pore volume and hence its ability to minimize coking.

Therefore, this study explores the production of liquid alkanes and cyclo-alkanes via the oligomerization of ethylene by loading 2.8wt % nickel on different SIRAL supports (SIRAL 10, SIRAL 20, SIRAL 30, SIRAL 40, SIRAL 40 HPV which have pore volumes of 0.75, 0.8, 0.9, 1.0 and 1.5 ml/g respectively) as SIRALs have a comparatively high level of surface acidity, large pore volume which minimizes coking making them useful as support materials. Testing the five catalysts synthesized (Ni-SIRAL 10, Ni-SIRAL-20, Ni-SIRAL 30, Ni-SIRAL 40, Ni-SIRAL 40 HPV) in the ethylene oligomerization reaction, this study will report the following (i) the activity of the different catalysts in the fixed-bed reactor used to carry out the oligomerization reaction. (ii) the most promising SIRAL support for increased liquid yield and product selectivity (iii) investigation of the effect of physiochemical properties obtained via characterization of the catalysts on the oligomerization reaction (iv) tested performance results of the Ni-SIRAL catalysts under different reactor conditions of varying temperature and pressure. Furthermore, the active metal (Ni) in the catalyst which produces the most promising results will be replaced with another active metal, chromium for further investigation.

111d Towards Developing a Pathway for Efficient Conversion of Syngas-Derived Olefins to Sustainable Aviation Fuel in High Yield and Selectivity

Anthony Giduthuri, Udishnu Sanyal, Mond F. Guo, Birgitte Ahring, Karthikeyan Ramasamy

Rapid efforts towards decarbonizing the aviation sector call for developing low-risk technologies that can easily be adapted to existing industrial processes, thereby improving the potential to be successfully commercialized. Though predominantly derived from fossil sources, syngas makes an attractive renewable feedstock to produce sustainable aviation fuel (SAF) from municipal solid waste (MSW) or any biomass via gasification that significantly reduces the carbon footprint [1, 2]. Various syngas-based pathways exist to produce aviation fuel, but bottlenecks such as low selectivity and/or yield to desired products and extensive downstream processing hinder these technologies. Syngas to methanol and methanol to olefins (MTO)[3] are current commercial processes with high process yields of 99% and 92%, respectively. Therefore, developing a concerted single-reactor process to integrate with MTO plays a crucial role in producing SAF from syngas using the olefins derived from the MTO process as intermediates.

Oligomerization of olefins, C_2 (ethylene), C_3 (propylene), C_4 (butene) and C_5 (pentene) have been investigated in the literature using heterogenous catalysis to produce fuel range hydrocarbons. However, there is a wide technological gap in the co-oligomerization of these olefins (specifically C2 and C3+ olefins), which typically is the makeup of the MTO process-derived olefins. The primary challenge is the difference in their reactivity and the corresponding oligomerization reactions between C_2 and C_{3+} olefins. Tuning different process parameters and catalyst composition has allowed us to obtain >90% conversion of C_2 and C_{3+} with about 80% selectivity to jet-fuel range hydrocarbons (C_8-C_{16}). We also tested the process at varied compositions of the olefin feedstock, which further helped us investigate the technology for adapting to various upstream processes that produce olefins. Most importantly, the catalyst developed here demonstrated ~150 hours of stable performance without deactivating.

1. Shahabuddin, M.; Alam, M. T.; Krishna, B. B.; Bhaskar, T.; Perkins, G., A review on the production of renewable aviation fuels from the gasification of biomass and residual wastes. *Bioresource Technology*, **2020**, *312*, 123596.
2. Centi, G.; Perathoner, S., Chemistry and energy beyond fossil fuels. A perspective view on the role of syngas from waste sources. *Today,* **2020**, *342*, 4.
3. Tian, P.; Wei, Y.; Ye, M.; Liu, Z. Methanol to Olefins (MTO): From Fundamentals to Commercialization. *ACS Catalysis* **2015**, *5* (3), 1922-1938. DOI: 10.1021/acscatal.5b00007.

111e Hydrothermal Liquefaction of the Brown Macroalgae *Saccharina Latissima* for the Production of Biofuel

Mayokun Oke, Fernando Resende

Seaweed aquaculture has matured over the last century for use as an energy source to produce biofuels which have a high feedstock requirement and need large amounts of biomass. Compared with land-based biomass, cultivated macroalgae in water presents numerous advantages such as freeing up land for food crops farming, increased carbon absorption, and cleaning of contaminated water. Macroalgae usually has a high-water content, as much as 90 wt. % in some species, and hydrothermal liquefaction (HTL) has gained attention due to its intrinsic advantage to produce biofuels using water in sub-critical region conditions as a reaction medium. This process eliminates the cost associated with setting up a separate drying unit for the feed, since the wet macroalgae can be used directly.

We performed hydrothermal liquefaction of brown macroalgae (*Saccharina latissima)* in the presence of homogenous and heterogenous catalysts such as KOH, Na_2CO_3, Pd/C, HZSM-5, Ce/HZSM-5, and Ni/SiO_2-Al_2O_3 in hot compressed water, to investigate their effects on product yields (biocrude, water-soluble product, gas, and solid residue) and compounds distribution in the biocrude. The experiments were conducted in a batch reactor at 330°C and 120-130 bar, for a residence time of 1 hour. To investigate hydrocarbon production, we also used methanol as a co-solvent (methanol-to-water 1/4 vol %) and as the only liquefaction solvent. We conducted further experiments on the macroalgae to investigate the effects of varying temperature, residence time and algae-to-water ratio on the product yields. Only one parameter was varied at a time for these catalytic hydrothermal liquefaction runs: at four different temperatures (240°C, 270°C, 300°C, and 330°C), three residence times (20 minutes, 40 minutes, and 60 minutes), and algae-to-water ratios (1:5, 2:15, 1:10 and 2:25). The final factor we studied was the effect of macroalgae pretreatment using deionized water on the product yield and biocrude composition. We conducted the gas Chromatography – Mass Spectrometry (GC-MS) analysis of the biocrude using Agilent GC 8890 equipped with a quadrupole mass spectrometric (MS) detector 5889, with a semi-polar capillary column. The GC inlet temperature was 300°C and the oven-programmed temperature was heated from 35°C to 300°C, at a heating rate of $10^oC.min^{-1}$. We identified the compounds in the biocrude using the National Institute of Standards and Technology Mass Spectra (NISTMS) library and quantified them by external calibration with analytical standards, to obtain their yields and selectivities. Finally, we explored pyrolysis as a secondary process to upgrade the biocrude obtained from HTL to produce hydrocarbons. We used a pyrolyzer connected to the GC-MS and conducted catalytic fast pyrolysis of the biocrude with HZSM-5 catalyst (1:1 and 1:5 weight-to-weight ratio).

The biocrude we obtained from all hydrothermal liquefaction runs was a very viscous dark-brown liquid with a pungent smell. The biocrude yields from the runs with water ranged from 12.9 wt.% (non-catalytic run) to 23.6 wt.% (with Ce/HZSM-5) and that of solid residue ranged from 5.0 wt.% (with KOH) to 19.1 wt.% (with Pd/C), while gas yields were from 18.6 wt.% (with KOH) to 28.3 wt.% (with Ni/SiO_2-Al_2O_3). The largest fraction of the products (23.8 wt.% - 39.9 wt.%) were dissolved organics in the aqueous phase. The maximum biocrude yield of 25.7 wt.% was obtained using methanol as a liquefaction co-solvent. We noticed no significant difference in the biocrudes yield from using methanol as co-solvent or only solvent, which could be because methanol can effectively extract and dissolve the organic matter from macroalgae regardless of whether it is used as co-solvent or only solvent. We analyzed the biocrude and found it to be a complex mixture of aldehydes, ketones, phenols, hydrocarbons, fatty-acids, esters, and heterocyclic nitrogenous compounds. Each catalyst led to varying selectivities for the classes of compounds identified in the biocrude. In the methanol runs, fatty acids and esters were the most abundant compounds identified in the biocrude with an increasing trend in their quantities for corresponding increase in the volume of methanol used as solvent. The best catalysts that showed relatively higher selectivities to produce hydrocarbons in water medium were HZSM-5 (3.8%), Ce/HZSM-5 (3.6%), and Pd/C (1.8%).

The highest average yield of biocrude (20.9 wt.%) was obtained at 330°C, revealing an upward tendency of yield with increasing reaction temperature. This assessment agrees with previous studies that have found that high temperature facilitates reactions such as repolymerization, hydrolysis and transformation in the hydrothermal liquefaction process. The biocrude yield also followed an upward trend for increasing residence time with 60 minutes leading to the highest yield in the range we studied. We found the effect of algae-to-water ratio on biocrude yield to be insignificant within the range of our study. Without pretreatment, the bio-crude yield was discovered to be lower, with higher yield of water-soluble products, and lower yield of gases and solid residues compared to the pretreated run. The yield of biocrude produced without pretreatment was 17.2 wt.%, compared to 20.9 wt.% after treatment. We hypothesized that sea salts present in the macroalga reduce the amounts of biocrude obtainable during conversion, because they can act as catalyst poisons during the HTL process. The biocrude we obtained without pretreatment, was mostly composed of complex nitrogenous compounds as well as unsaturated ketones and ethers, compared to the pretreated macroalgae under the same operation condition which had a high selectivity for fatty acids & esters. Upgrading of the biocrude through pyrolysis-GCMS led to very high hydrocarbon selectivity of about 90%, majorly composed of aromatics like benzene, toluene, styrene, and ethylbenzene.

These results show promise in assessing the prospect of using *S. latissima* as a natural source to produce transportation fuels and high-value chemicals. All the catalysts led to increased biocrude yields, and the choice of catalysts and solvent had significant impact on the composition of biocrude produced during the liquefaction process. Temperature was discovered to be the most important factor affecting the biocrude yield, with the amount of biocrude obtained increasing from 11.1 wt.% (at 240°C) to 20.9 wt.% (at 330°C). Of the three tested residence times, 60 minutes was most favorable as it allowed enough time for the maximum amount of biocrude (20.9 wt.%) to form without losing components to the water-soluble products or solid residues. The effect of algae-to-water ratio was not very pronounced as the variation in biocrude yield was minimal. Pretreating of the feedstock by washing with deionized water generally helped in eliminating impurities, improving biocrude yield and the distribution of compounds in the obtained biocrude. Pyrolysis appeared to be a viable process for upgrading this biocrude to more closely approximate transportation fuels and obtaining high-value chemicals. The zeolite catalyst (HZSM-5) was most effective in the biomass-to-catalyst ratio, 1:5 to pyrolyze the biocrude obtained from hydrothermal liquefaction to produce 4.1 wt.% hydrocarbons (89.8% selectivity), that was primarily composed of aromatics.

111f Separation of Torrefaction Bio-Oils Using Ammonia

Samuel Drummond, Thossaporn Onsree, Alexis Allegro, Jochen Lauterbach

In order to decrease overall carbon emissions, biomass related fuels have developed a renewed interest. One option is to combust biomass, however in order to allow for use in current electricity generating systems, pretreatment is required. Raw biomass contains water and other non-preferable chemicals for combustion which can be removed using torrefaction or pyrolysis. The resulting bio-oils create other issues, though, as their large water content and acid content make storage and combustion a problem[1]. Our process uses ammonia to create usable mixtures for these bio-oils. Here, ammonia is bubbled through the liquid, which dissolves in the water and neutralizes the acidic components, before separating the bio-oil into an aqueous phase and non-aqueous phase. This was tested on a model bio-oil first, and the water based layer contained a relatively pure solution of ammonium acetate. The organic phase, however, contained many heterocyclic compounds, making usage as a feedstock difficult. This bio-crude was easily combustible, with a heating value slightly higher than ethanol. This process was then repeated using wood pellets as a feedstock for the bio-oil. Separation occurred once again with the aqueous phase being tested as a possible fertilizer, in an attempt to close a loop of nutrients for plant growth and usage as an energy source. The organic phase contained mostly long chain compounds, and cracking of this bio-crude was conducted using H-ZSM 5 catalysts. This was tested over a number of conditions and yielded a mixture of benzene, toluene, xylenes, acetonitrile, furan derivatives and phenol derivatives, which could be used as a fuel. Overall this process was shown to create value added products from a previously unusable chemical mix.

1. Yang, H; et al. Overview of Upgrading of Pyrolysis Oil of Biomass. *Energy Procedia*. **2014**, 61, 1306-1309.

111g Bio-Jet and Bio-Kerosene Production from Ozonolysis of High Oleic Oil

Junli Liu, Nathan S. Mosier, Elena Robles Molina, Xueli Chen, Gozdem Kilaz, Louis Edwards Caceres Martinez

United States must reduce greenhouse gas (GHG) emissions by at least 50 % below 2005 levels by 2030 and achieve net zero emissions by 2050. The transportation sector accounts for over 25 % of total GHG emissions in the United States. With the action of net-zero emission, passenger vehicles and light-duty diesel engine trucks will be phased out by electric vehicles by 2035. However, heavy-duty engines, such as airplanes, will still require liquid fuels with a projected annual production of 36 billion gallons, and over 50 % of jet fuels need to be renewable. Several technologies have been developed to produce fuels from renewable sources to fulfill net zero emission, such as hydrogenation of fats/oils, alcohol dehydration/oligomerization, sugar hydrogenation, syngas hydrogenation, and biodiesel fractionation. However, these technologies have high energy requirements for production, low yields of target products, or high energy needs for product separation. High oleic oils initially have received attention for their health and nutrition benefits, but industrial applications are now receiving attention. The production of high oleic soybean oil will significantly increase from 3 billion pounds in 2022 to 9 billion pounds in 2027. In this study, nonanoic acid from ozonolysis of oleic acid in high oleic soybean oil was used as the feedstock to produce various alcohol esters via methanesulfonic acid catalysis. Varying the carbon length of the alcohol exhibits an insignificant effect on the density of the products but significantly influences other properties, such as cloud points, viscosity, and heat of combustion. Our results show that these products could be excellent alternatives to jet fuels or kerosene because of the excellent cold flow properties with the cloud points ranging from -35 ^{o}C to - 70 ^{o}C depending on the type of alcohol. The other properties, such as density, viscosity, energy contents, cetane number, etc., were also measured by ASTM standards and complied with the ASTM standards. In addition, the properties can be adjusted by the type of alcohol for various use. Therefore, the approach we described provides a novel method to produce alternatives to petroleum-based jet, kerosene, and diesel fuels.

111h Electrocatalytic Conversion of Phenol to Cyclohexane

Jeffrey Page, Amol Pophali, Tae Jin Kim, Stoyan Bliznakov, Julia A. Valla

The production of renewable fuels has become necessary as fossil fuels continue to deplete past the point of no return.[1] Pyrolysis of biomass to liquid fuels has emerged as a promising solution due to the possibility of conversion of high carbon content feeds into fuel. Pyrolysis oils, however, struggle from low yields and increased heteroatom content.[2] In order to solve this problem a second upgrading step is needed in order to increase the fuel's hydrogen content and remove heteroatoms. Conventional oil upgrading has been done through thermochemical processes using high temperatures (>300 °C) and high-pressure hydrogen (10-20MPa).[3,4] Electrochemical upgrading of bio-oils represents a novel avenue for the conversion of bio-oils to usable fuels or commodity chemicals under mild condition which requires milder temperatures (30-80 °C) and generates the hydrogen *in situ* through the water splitting reaction.[4] Though initial results on electrocatalytic upgrading have been promising few reports appreciable yields of fully deoxygenated products and the efficiency remains low at high current densities.[4]

This work presents the upgrading of phenol as a bio-oil model compound. ECH of phenol was investigated in a custom electrochemical cell (H-Cell), wherein the anode and cathode are separated by a Nafion® 117 membrane. Preliminary ECH results using a commercial PtRu-C catalyst showed that after 4 hours of electrolysis at 55 mA cm^{-2}, phenol conversion reaches 80%. The primary products of phenol ECH are cyclohexanol and cyclohexane with selectivity of 59% and 27%, respectively. Further ECH experiments were performed using lab synthesized PtRu, Pt and Ru catalysts to investigate the role of active metal on cyclohexane selectivity. These results (figure 1b) showed that the lab synthesized PtRu catalyst produced the highest cyclohexane selectivities (31%), highest conversion (100%) and efficiencies (35%). ECH of phenol was further investigated using operando Raman spectroscopy to understand the mechanism of cyclohexane formation. Preliminary results from these experiments showed that the cleavage of the C-OH bond depends on the potential.

(1) Zhang, X.; Lei, H.; Chen, S.; Wu, J. Catalytic Co-Pyrolysis of Lignocellulosic Biomass with Polymers: A Critical Review. *Green Chemistry*. 2016. https://doi.org/10.1039/c6gc00911e.

(2) Zhang, B.; Zhong, Z.; Min, M.; Ding, K.; Xie, Q.; Ruan, R. Catalytic Fast Co-Pyrolysis of Biomass and Food Waste to Produce Aromatics: Analytical Py-GC/MS Study. *Bioresour. Technol.* **2015**, *189*. https://doi.org/10.1016/j.biortech.2015.03.092.

(3) Page, J. R.; Manfredi, Z.; Bliznakov, S.; Valla, J. A. Recent Progress in Electrochemical Upgrading of Bio-Oil Model Compounds and Bio-Oils to Renewable Fuels and Platform Chemicals. *Materials (Basel).* **2023**, *16* (1), 1–33.

(4) Chen, G.; Liang, L.; Li, N.; Lu, X.; Yan, B.; Cheng, Z. Upgrading of Bio-Oil Model Compounds and Bio-Crude into Biofuel by Electrocatalysis: A Review. *ChemSusChem*. 2021. https://doi.org/10.1002/cssc.202002063.

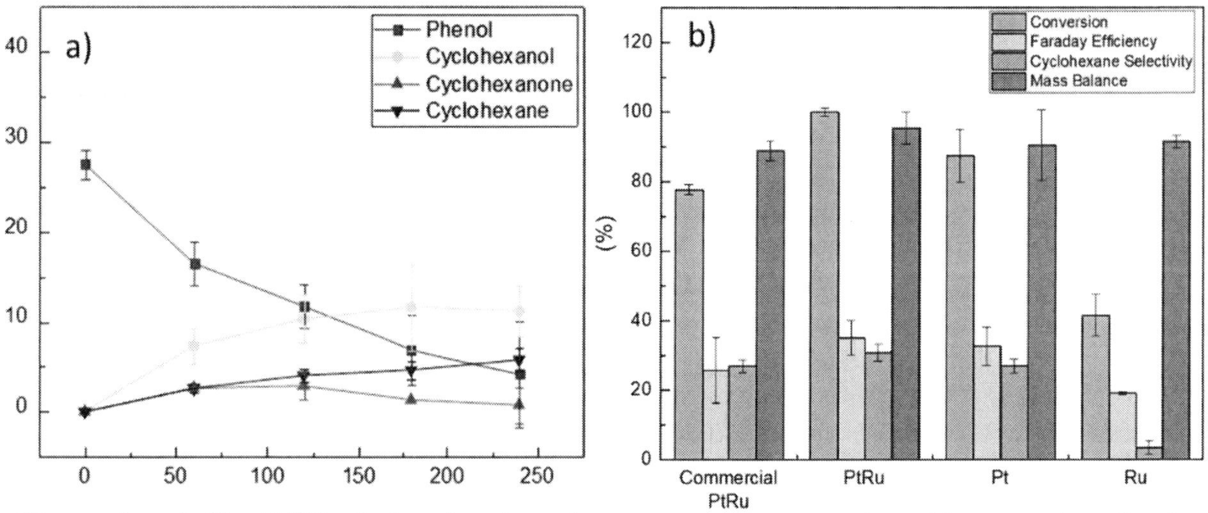

Figure 1. a) The ECH of phenol and product concentrations vs time. b) conversion, faraday efficiency and cyclohexane selectivity for phenol ECH with different catalysts. Reaction conditions: 55 mA cm^{-2}, 50 mL min^{-1} N_2, 60 °C, 0.2 M perchloric acid, 0.2 mg cm^{-2} catalyst loading.

111i Assessing the Long-Term Stability and Degradation of a Marine Advanced Biofuel Blend

Patritsia M. Stathatou

The shipping sector contributes 3% to annual global greenhouse gas (GHG) emissions. With the sector's continual expansion, stringent environmental regulations are being enforced, compelling ship owners to minimize emissions. Biofuels produced from renewable feedstocks are deemed a highly viable option to achieve decarbonization targets in the short-term future. Specifically, advanced or second-generation biofuels, derived from non-food biomass, wastes, and residues, are considered the most sustainable and readily available biofuel option. These biofuels can be blended with conventional marine fuels and used in existing engines with minimal or no adjustments (drop-in fuels), leveraging existing bunkering infrastructure. Our prior investigation examined the potential of advanced biofuel blends to decarbonize dry bulk shipping, verifying their efficient burning and operational safety. Nevertheless, potential long-term consequences of utilizing and storing these biofuel blends onboard have not been researched, creating uncertainty concerning proper engine performance and operation.

In this presentation, the results of a comprehensive storage and testing protocol will be discussed, developed to investigate the stability and degradation of a 20% biofuel blend of used cooking oil (UCO) biodiesel and very low sulfur fuel oil (VLSFO) without the use of any antioxidants over a period of 12 months. Three sample groups, each comprising five 1L samples, have been assigned and stored under different temperatures, *i.e.*, 4°C, 23°C and 50°C. In each group, five different storage configurations have been designed to account for various environmental conditions, such as water presence, light and oxygen exposure. Eight parameters have been monitored in all samples for one year following standardized procedures, *i.e.*, acid value, microbial contamination, density, viscosity, total sediment potential, water and sediment, peroxide value and oxidation onset temperature. Our results indicate that minor degradation and microbial contamination were evident after the first three months of storage, and gradually worsened over time, particularly in samples with water and oxygen exposure. Potential impacts of such degradation on fuel properties and engine operation will be discussed together with relevant mitigation actions, providing valuable insights to both biofuel producers and users for maintaining fuel quality and ensuring long-term engine performance.

252a Introduction to the Future of Petroleum Industry

M. R. Riazi, Vladimir Mahalec

The petroleum industry is going through a challenging period with strict economical and environmental constraints. It is important to understand the future of crude oil production and refining and how the industry would survive under different scenarios. Today's technology choices and the upgrading of petrochemical facilities, however, need to be driven by a completely different logic than the existing industry. Current operations were built on the basis that economics of carbon dioxide are only due to fuel cost, and that the environmental aspects of fuel use are dominated by sulfur emissions, nitrous oxides and particulates. The future of regulations and market acceptance look very different in the future in many countries, and these changes are driving huge financial decisions in the refining industry.

252b Measurement of Calcite Solubility at Temperatures to 573K and Pressures to 68 Mpa

Hseen O. Baled, James Fazio, Ward A. Burgess, Isaac Gamwo

Calcite precipitation during oil production can pose a serious hindrance to the oil flow. Yet, calcite solubility experimental data needed for the prediction of calcite precipitation in ultradeep offshore reservoirs are not available in the open literature at temperatures above 523K.
The National Energy Technology Laboratory (NETL) recently designed, built, and commissioned a new high-temperature, high-pressure solubility apparatus rated to 600K and 68 MPa. The new apparatus was tested for its reliability by measuring calcite solubility at conditions available in the literature. The apparatus was then used to extend the literature calcite solubility data to high temperatures up to 573 K and pressures to 68 MPa in a high-salinity brine containing NaCl and Na_2SO_4. Such data are needed to experimentally validate thermodynamic models being developed to predict and inhibit the formation of calcite during oil production.

252c Providing Guidelines for Thermal Removal of Paraffin Deposits from Production Lines

Elijah Bell, Gabriel Santos, Nagu Daraboina, Cem Sarica

Wax deposition in subsea crude oil pipelines is a major operational issue in oil production and transportation. Several methods are used in the prevention and remediation of wax deposition, including pigging, chemical inhibition, thermal insulation, and active heating. Active heating involves using a heating medium or electrical current to heat the wall of the pipe carrying the production fluid and is increasingly used as a cost-effective strategy for controlling wax deposition in offshore flowlines. While these techniques have been widely used, there exists no clear understanding of the mechanism and guidelines for wax removal using active heating technology. Currently, operators tend to aim for temperatures far above the WAT to ensure removal. There is an interest in finding actual temperature targets for wax removal. If these can be defined and current operational targets are lowered, this will reduce energy requirements and therefore reduce costs. To this end, a novel dynamic microscopic visualization technique is employed to understand the removal mechanism as the deposit is exposed to varying removal temperatures.

A mechanism for thermal removal is proposed based on visual evidence for a counter-diffusion process preceding detachment. During removal, the deposit becomes less dense with wax crystals over time, lowering its yield stress until a point where shear forces from the flow can strip and carry away large chunks of wax deposit. A method for determining the minimum temperature of removal is proposed based on this mechanism and validated using yield stress correlations based on an in-depth rheological study of waxy oils. The pipe wall needs to be raised at least to this temperature for removal to occur. The higher the wall temperature is raised above this value, the faster the removal will occur. Understanding this mechanism can inform the development of accurate removal models and improve production economics.

252d Automated in-Line Optical Trace Level Detection of Expanded Palette of Advanced Tracers

Sehoon Chang, Gawain Thomas, Wei Wang

Oil field tracer technology development to illuminate reservoir characteristics such as well connectivity, volumetric sweep efficiency, and geological heterogeneity for the purpose of improving oil production optimization has gained momentum over the last decade. In this study, we demonstrate the laboratory qualification of an expanded number of ultra-trace level detectable molecular isomers as competent cross-well water tracers for use in highly retentive carbonate reservoirs with harsh salinity and temperature. Our Advanced Tracers technology has been using time-gated optical fluorescence with high sensitivity, selectivity, and oil blind background, by chelating the barcoded variants of the ligand tracer molecules with lanthanide ions. The ultra-trace level detection of expanded palette of barcoded molecular isomer tracers has been achieved under flow conditions utilizing a microfluidic mixing chip and a flow-cell fluorescence or UV-Vis detector at the output of the high-performance liquid chromatography (HPLC) separation. This study has demonstrated efficient separation and rapid in-line detection of a mixture of the expanded palette of molecular isomer tracers at ultra-trace level. HPLC chromatographic separation equipped with automated solid phase extraction (SPE) pre-concentration for molecular isomer tracers has lowered the in-line fluorescence detection limit down to 500 pM of molecular isomer tracers. The developed complete automated tracer detection system will offer unprecedented sensitivity of detection and accelerate the sample analysis aiming "near-real-time" oil field tracer analysis. The developed HPLC chromatographic separation and optical detection of isomers of molecular tracers can significantly expand the palette of barcoding capability. The automated in-line optical detection technology for tracers in oil field brine will pave the way toward the development of a near-real-time tracer sample analysis approach for wide full-field applications.

252e Nanotags for Improved Cutting Depth Correlation

S. Sherry Zhu, Marta Antoniv, Gawain Thomas, Martin Poitzsch, Hyung Kwak

Real-time evaluation of drill cuttings in the field is highly desired, as direct petrophysical characterization of the formation can effectively support logging, measuring while drilling (LWD and MWD), and geosteering when integrated into mud logging workflows. This direct characterization of drill cuttings, performed while drilling and integrated into basic mud-logging workflows, provides an alternative to costly well-logging. Advanced mud logging is particularly promising for coiled tubing drilling, where advanced LWD tools may not be available due to the reduced hole size. In unconventional reservoirs and for extended-reach horizontal drilling circumstances, an economically affordable formation evaluation solution can potentially optimize well placement, improve efficiency and safety of drilling and completion, and ultimately maximize recovery cost-effectively.

Despite its potential benefits, advanced mud logging deployment has historically been plagued by poor data quality due to uncertainties in drill-cutting depth determination. The accuracy of cuttings' depth assessment depends on a number of factors, including wellbore mud hydraulics, hole cleaning, and an accurate lag time of the cuttings. However, the lag time can be difficult to determine accurately, as it is based on the cuttings' return trip up the annulus and may be affected by systematic overgauge drilling, hole caving, or spiraling factors that are not accounted for in the mud-logger's software. Inaccuracies can also arise from gravitational debris accumulation, hydraulics, and hole cleaning, especially in long horizontal sections. When the cuttings' return trip lasts half an hour or longer, depth uncertainties can be significant, ranging from ten to twenty feet or more. Furthermore, inaccurate labeling of collected cuttings or mis-sorting during handling can exacerbate these errors.

To address these challenges, we have developed a groundbreaking technology that enables accurate tagging of drill cuttings using barcoded NanoTags during drilling operations. This innovative depth correlation system comprises three components: barcoded NanoTags, an extraction process, and a detection method that can precisely and uniquely detect the tags present on the drill cuttings. The tags are injected into the water-based mud at ppm level as the tag enters the drillstring and label the freshly generated cuttings at the drillbit. The tagged cuttings are collected over the shale shaker at the estimated tags' arrival time, and the lag time of the tags is determined by their downward-trip time through the dimensionally accurate drill pipe or coil ID. This method significantly reduces depth-correlation uncertainty to +/- 2 ft compared to conventional method that estimates the lag time using the cutting's return trip up the annulus.

The NanoTags are custom-polymerized nanoparticles that range in size from 20-60 nm. The NanoTags, synthesized using an emulsion polymerization, are compatible with water-based drilling fluids, stable in subsurface conditions, and have a shelf life of more than one year at ambient conditions. Some NanoTag formulations also contain a fluorescent dye. The tags are isolated from the cuttings in the lab using a workflow comprised of water washing, solvent extraction, and thermal separation. Since the NanoTags generate unique mass spectral signals, we can detect them unambiguously at μg levels using a GC/MS equipped with a pyrolyzer (Py-GCMS). The fluorescent tags on cuttings or in mud are visible under a portable fluorescence microscope.

Furthermore, we have customized a robotic platform, built by Chemspeed Technologies, to process and prepare the field cuttings for analysis. The Chemspeed platform can weigh cuttings, add solvent, filter the cuttings, and vacuum dry 16 different types of cuttings, or 48 samples total when measured in triplicate, within 24 hours. These steps recover the tags from the field cuttings for final analysis using pyrolysis-GCMS. The customized automation platform significantly reduces the time and manpower necessary to prepare the cuttings for characterization.

We have demonstrated the effectiveness of the NanoTagging technology during two field tests performed in 2019 and 2022 in Saudi Arabia during the drilling of two different vertical gas wells. Both field tests confirm the successful tagging of the drill cuttings with our tags and show the improved depth correlation of the cuttings. Furthermore, we designed an air-powered rapid-multiplexing tags injection apparatus to deliver sharp pulsed injection of each tag continuously to the mud during the 2nd field test using pre-programmed injection times and flow rates for each tag.

The NanoTags for drill-cuttings depth correlation technology improves the current mud logging process and is cost-effective, straightforward, and environmentally friendly. This technology allows oil and gas operators to improve the quality of lithological and petrophysical analyses, enhance formation evaluation and geosteering practices, and control well placement with mud logging, thus helping to maximize oil recovery. The automation of the field dispensing of the different NanoTag barcodes into the mud and the automation of the lab extraction and detection workflow in combination with optical detection methods could make the tagging process a routine operation available for improving mud-logging data quality in the field.

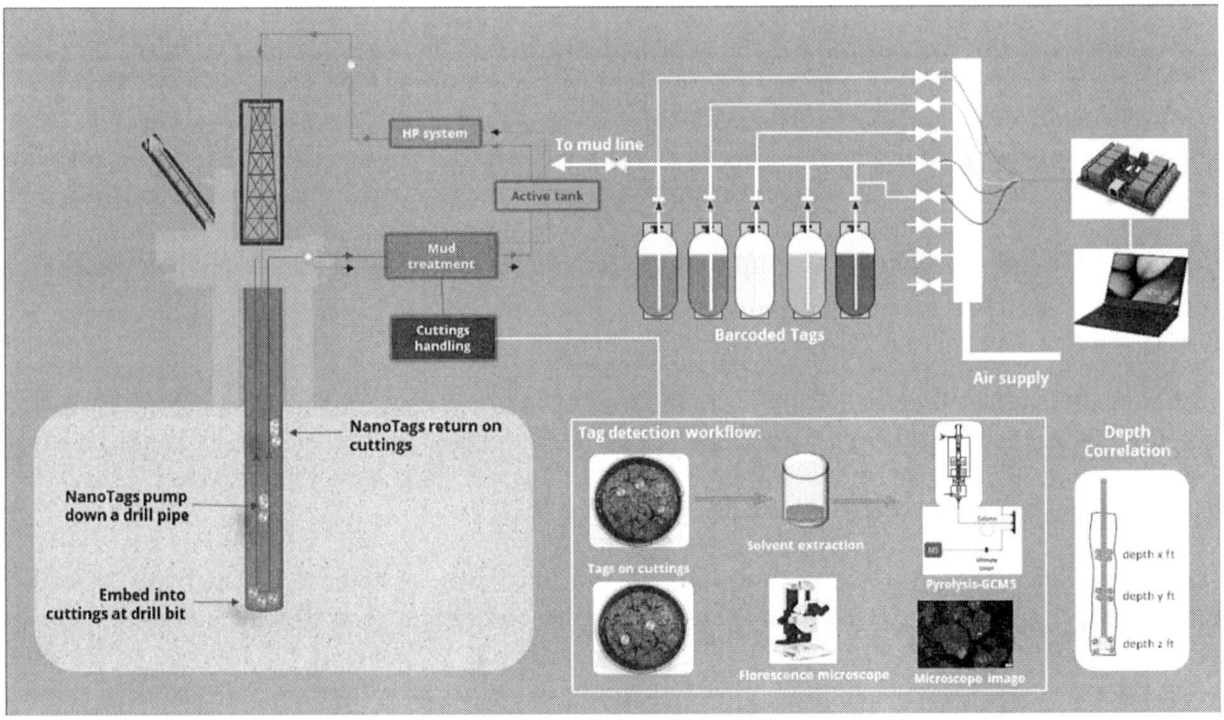

252f A Rapid Microwave-Assisted Bottle Test Method to Determine the Crude Emulsion Stability

Prasad Pawar, Clayton Jeffryes

The main objective of this study is to use microwaves to develop a fast, energy efficient, and economical method to quantify the stability of crude oil and water emulsions. The crude oil emulsions are formed by high shear mixing of process water and crude oil during pumping, desalting and transportation through wellbores and pipelines and are stabilized by emulsifying agents that form interfacial films at the oil-water interface. Separation of these crude oil emulsions is expensive and requires significant amounts of energy and chemical demulsifiers. Due to the substantial cost of chemical demulsifiers used during the crude processing, there is a commercial interest in developing an inexpensive method to optimize their use. Currently, the widely used industry standard for this measurement, the Bottle Test, can be time consuming and has limitations when dealing with extremely heavy crude emulsions that sediment over long timescales because of low density differences and also with tight emulsions that flocculate and coalesce over long timescales because of high interfacial area to volume ratios. Also, the said Bottle Test is performed off the oil fields, usually in a third-party laboratory. Thus, increasing time and cost to perform the optimization process. In this study, we have created a rapid and simple microwave demulsification test that would be used in correlation to the standard Bottle Test.

The tight crude oil emulsions, characterized by suspended droplets on the order of submicron to tens of microns, often cause serious problems to the wellbores and pipelines such as clogging, corrosion, and pump failures. Our previously developed microwave demulsification process proved to be orders of magnitude faster at separating heavy crudes and tight emulsions when compared to heat and gravity settling alone, as used by the traditional Bottle Test.

To simulate field conditions, we mixed two crude oils to prepare samples with American Petroleum Institute (API) gravities between 28.9° and 46.0°. These crudes were mixed with water to create emulsions containing between 20% and 80% water. This array of Water-in-Oil (W/O) and Oil-in-Water (O/W) emulsions with different densities and viscosities were mixed at varying shear rates. A Polyol Block copolymer, and Polyethylene Glycol 2000 (Mol. Wt. 2000) were used as demulsifying agents at different concentrations. The crude oil emulsions mixed with the demulsifying agents were subjected to both the traditional Bottle Test as well as the rapid microwave test. The amount of phase separation was visually recorded and the residual water content in the separated oil phase was measured using the Karl Fisher titration. Using the separation data for both methods, we developed a correlation between the two methods. The results from the rapid microwave test can now be used to determine how they correlate with the standard bottle test.

252g Integration of Matrix Encoding in Transformer Model for Reaction Pathway Prediction in Oil Refining Process

Qiong Pan, Jie Li

Due to the foreseeable shortage of fossil fuels, increasing global energy demand, and growing environmental concerns, effective utilization of fossil fuels with the concept of 'molecular management' has attracted increasing attention from researchers. Molecular management targets the right molecules to be at the right place, at the right time and at the right price. Molecular management involves a range of techniques, including molecular characterisation of refining streams, molecular modelling to optimise refining processes, and integration of processing and utility systems at the molecular level, to achieve overall refinery optimisation. By leveraging these advanced techniques, it is possible to maximize the efficiency of fossil fuel usage while minimizing its impact on the environment.

In our last works, we developed a method that automatically converts molecular structure (represented using SMILES, simplified molecular-input line-entry system) to a list of submatrices that represents the connectivity information of each substructure in the molecule, then structure-property relationship model was then developed using neural networks. Despite the molecular property prediction model, a molecule composition reconstruction model that was developed. The surrogate composition model using SMILES as basic representation unit, homologous series of molecules is generated by configuring the sidechain structure to the core structures. To ensure molecular diversity, a random distribution is used, while to reduce the degree of freedom of the model, a probability density distribution is adopted. A hybrid optimization algorithm consist of genetic algorithm and Sequential Least Squares Programming (SLSQP) algorithm is used to increase the numerical robustness and reproducibility of the results. The composition reconstruction model was evaluated by applying it in two case studies including petroleum fractions and biomass pyrolysis oil.

The Transformer model is a type of neural network architecture that can process sequences of input data, such as natural language sentences or audio signals, and produce corresponding outputs. It has shown significant improvements in many natural language processing tasks, such as machine translation. Chemical reaction prediction can also be treated as a machine translation problem between SMILES strings of reactants, and products, which has been investigated in the past few years.[1-5] However, the encoding of molecular structures is different to natural language, the SMILES string structure representation can be difficult to interpret, and sometimes the generated SMILES strings of products can not be canonicalized. Therefore, we tried to integrate the structural submatrices representation method into the transformer reaction prediction model to improve production accuracy. A reaction database covers petroleum refining process including cracking, substitution, Isomerization, etc., is used to develop the reaction prediction model. The results show the prediction accuracy has increased significantly over the traditional template-based methods. In addition, the generalizability of the transformer model greatly simplifies the process.

References

[1] D.M. Lowe, Extraction of chemical structures and reactions from the literature, University of Cambridge, 2012.

[2] C.W. Coley, R. Barzilay, T.S. Jaakkola, W.H. Green, K.F. Jensen, Prediction of Organic Reaction Outcomes Using Machine Learning, ACS Central Science 3(5) (2017) 434-443.

[3] P. Schwaller, T. Laino, T. Gaudin, P. Bolgar, C.A. Hunter, C. Bekas, A.A. Lee, Molecular

Transformer: A Model for Uncertainty-Calibrated Chemical Reaction Prediction, ACS Central Science 5(9) (2019) 1572-1583.

[4] M.H.S. Segler, M.P. Waller, Neural-Symbolic Machine Learning for Retrosynthesis and Reaction Prediction, Chemistry – A European Journal 23(25) (2017) 5966-5971.

[5] P.P. Plehiers, C.W. Coley, H. Gao, F.H. Vermeire, M.R. Dobbelaere, C.V. Stevens, K.M. Van Geem, W.H. Green, Artificial Intelligence for Computer-Aided Synthesis In Flow: Analysis and Selection of Reaction Components, Frontiers in Chemical Engineering 2 (2020).

252h Hydroxybetaine Zwitterionic Surfactants for Enhanced Oil Recovery: Compatibility with Acrylamide-based Copolymers and Rheological Properties

Muhammad Shahzad Kamal, Afeez Gbadamosi, Shirish Patil, Syed Hussain

Oil is extracted from a reservoir in a three-stage process. During the initial and secondary stages, oil is extracted through the reservoir's natural pressure or by water injection, but this method only recovers 10% to 25% of the total oil present. Therefore, tertiary or enhanced oil recovery (EOR) methods are used to retrieve the remaining oil. Chemical methods are used in EOR, where various chemicals such as surfactants, polymers, and alkalis are injected to increase hydrocarbon recovery. Polymers are utilized to increase water viscosity, improving the oil/water mobility ratio, whereas surfactants are utilized to alter the interfacial tension between water and oil. In most cases, surfactants and polymers are injected together to achieve better mobility and lower interfacial tension, which is a challenging task due to harsh reservoir conditions. The research work under discussion synthesized polymeric hydroxybetaine zwitterionic surfactants, which were used with acrylamide-based copolymers that are thermally stable and salt-tolerant. The rheological properties of the surfactant with copolymer were analyzed using an Anton Paar rheometer, indicating compatibility at high temperatures with no precipitation. The surfactant reduced the interfacial tension between oil and water and demonstrated stable rheological properties with a stable viscosity profile over several days. The elastic modulus was also stable. The viscosity of the polymer decreased with the addition of surfactant in deionized water due to charge screening, but the viscosity of the polymer remained unaffected in seawater. The oil recovery experiments using the developed surfactant polymer system showed more than 15% oil recovery.

252i Interfacial Behavior of Anionic Surfactant, Biopolymer and Hydrophilic Nanoparticles with Heavy Crude Oil for Enhanced Oil Recovery Applications in Sandstone Reservoirs

Rahul Saha

The demand of crude oil exploration from matured reservoirs is inevitable due to world energy demand. As the light category crude oil (easy oil) is exhausting, therefore the extraction of heavy crude oil from sandstone reservoirs is of prime focus. Thus, to enhance heavy residual oil recovery, nanofluids (surfactant-polymer-nanoparticles) flooding enhanced oil recovery schemes were investigated. Nanofluid flooding is an emerging trend in oil and gas industry however, their successful implementation in heavy oil reservoir is challenging due to severe complexities of the oil reservoirs system. Till today, no commercial successful nanofluid flooding has been executed or deployed in commercial oil field scale. Therefore, the existing condition conveys the importance of nanotechnology for oil recovery applications. In this work, a nanofluid formulation was developed with an understanding to recovery heavy oil from oil reservoirs. An initial screening of surfactant and polymers were executed based on the interfacial and rheological properties respectively. Numerous surfactants from anionic and non-ionic categories were tested and the surfactant which retains lower interfacial tension (IFT) along with emulsion formation capabilities with the heavy oil was selected for sandstone reservoir. Anionic surfactant (Triton X 100) showed lowest IFT behavior by reducing the IFT by almost 200 times. In addition, stable emulsion behavior was observed with Triton X 100 as compare to other surfactants which has the potential to improve oil displacement efficiency. Additionally, as Triton X 100 is non-ionic it can be deployed to reservoirs of any charge and is expected to more or less similar adsorption characteristics. Similarly, biopolymer Xanthan gum was considered based on its tolerant behavior for high saline and high temperature so that it can retain its rheological properties irrespective of extreme reservoir conditions. Nanoparticles SiO_2 (hydrophilic) was chosen based on its ecofriendly nature and is also one of the commonly used nanoparticles. The dispersion stability of nanofluid solutions were examined by sedimentation, particle size analyzer and zeta potential analysis and further inspection for mechanistic investigation were carried out. The nanoparticles were unstable in aqueous phase and show more than 75% sedimentation in just one day however, in aqueous chemical (surfactant –polymer) phase the nanoparticles were stable even after a week. The stability was further confirmed by negligible variation in average particles size value and zeta value of above ±30. The mechanisms as interfacial tension, interfacial packing of surfactant-polymer-nanoparticles, interfacial properties, emulsion, droplet size distribution, creaming rate, rheological properties, wettability change, displacement efficiency and flow of nanofluid in porous media of reservoir rock were scrutinized in detail. The superior interfacial tension reduction (24 mN/m to 3 mN/m), emulsion droplet size reduction (7 to 5 μ) confirming stable emulsion, lower creaming rate (Creaming Index ≥ 80), viscosity enhancement (450 to 650 mPa s), drastic contact angle reduction (75° to 14°) due to higher disjoining pressure, and improved oil displacement validated by flooding experiments were all confirmed as optimum concentration of nanoparticles in the nanofluid solution was reached. Finally, core flooding experiments in sandstone berea cores resulted in an additional oil recovery of 28% at 30°C and 24% at 80°C. The higher pressure drop curve during flooding analysis indicated the formulation of oil bank and its corresponding oil recovery with additional adsorption of nanoparticles on reservoir rock pores.

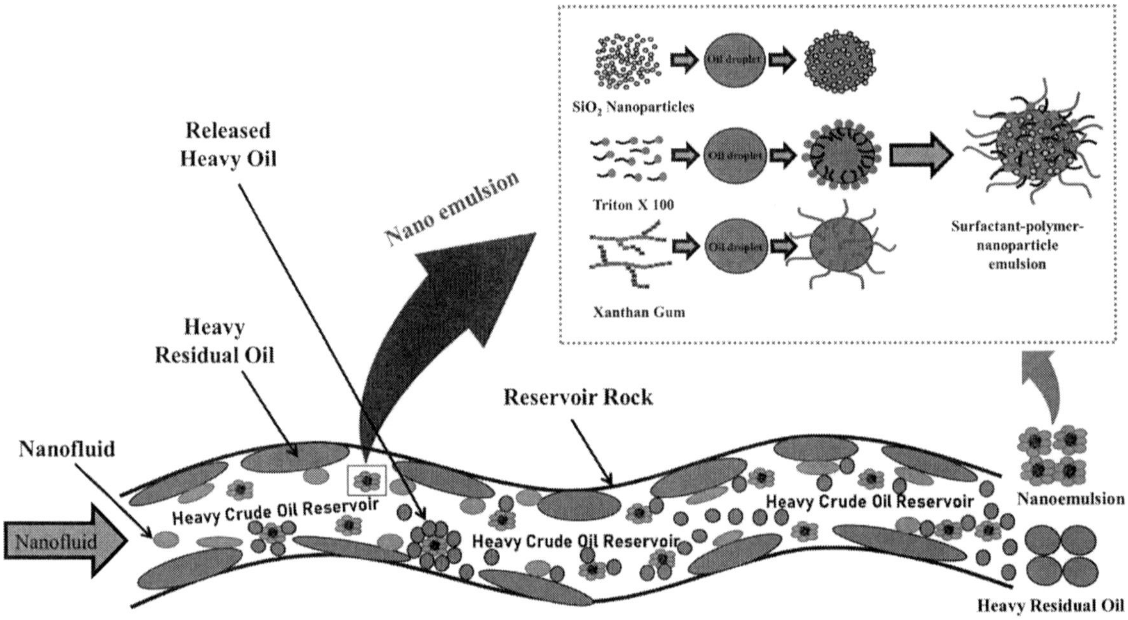

SiO2 Nanoparticles
Oil droplet
Triton X 100
Xanthan Gum
Surfactant-polymer-nanoparticle emulsion
Released Heavy Oil
Nano emulsion
Heavy Residual Oil
Reservoir Rock
Nanofluid
Heavy Crude Oil Reservoir
Heavy Crude Oil Reservoir
Heavy Crude Oil Reservoir
Nanofluid
Nanoemulsion
Heavy Residual Oil

343a Decarbonization of Long-Haul Trucking: LOHC Powertrains

Sayandeep Biswas, Kariana Moreno-Sader, Rob Jones, William Green

Decarbonization of trucks is a necessity to achieve 2050 climate goals. Trucking has been classified as a tough to decarbonize sector by International Energy Agency (IEA). Their report forecasts the sector's emissions to be 2.5 Gigatons of CO_2 per year by 2030, a 25% increase from 2020. Long-haul trucks make up 9% of the global cargo vehicles but account for 39% of the sector's greenhouse gas (GHG) emissions. In the US, long-haul trucks presently carry 71% of the payload making them an integral part of the transportation network. The demand for long-haul trucks is expected to double between 2020 and 2050, making it critical to find renewable powertrains to replace diesel trucks. There are numerous decarbonization routes such as biodiesel, battery electric, and hydrogen-based options. However, there is no silver bullet, as each option has its own set of benefits and drawbacks. Decarbonization decisions require an accurate assessment of all trucking options which allows us to make apples-to-apples comparisons.

Previous work by our group, has developed a comprehensive framework to compare the total cost to society (TCS) and well-to-wheel emissions (WTW) associated with each powertrain option. The powertrains are evaluated on an in-house drive cycle developed from 58,000 miles of real-driving data and represents long-haul trucking in the US. The commercial software GT-Suite is used to perform all the vehicle level simulations to obtain fuel economies and tail-pipe emissions for the different powertrain options. The fuel economy is used as an input to a techno-economic model to calculate the TCS for the powertrain. While the tail-pipe emissions become an input to the detailed emissions analysis to get the WTW emissions associated with each powertrain option.

Applying this framework to hydrogen (cryogenic liquid, compressed gas) powertrains we found that the TCS is about a factor of two higher than the TCS shipping via diesel trucks in the short-term scenario. Even for the long-term scenario, we find that the TCS for hydrogen trucks is 25% more than diesel trucks. A large portion of this cost is associated with the cost of hydrogen at the retail station, which is dominated by the cost of transportation and refuelling, not H_2 production. Traditionally, hydrogen is transported either as a cryogenic liquid or a compressed gas. Liquefaction and compression are both energy-intensive processes. Additionally, existing infrastructure is unsuitable for cryogenic liquids or compressed gasses, making it a significant expense.

A promising option to address the current pain points with hydrogen powertrains is the use of liquid organic hydrogen carriers (LOHCs). LOHCs benefit from being non-reactive, stable, room temperature liquids that are easy to transport and store hydrogen within their chemical bonds. This allows considerable reductions in transportation costs as it benefits from the synergy with the existing fuel transportation infrastructure. However, liberating at H2 from LOHC at the refuelling station (to fuel hydrogen trucks) has drawbacks: First, there is a 30% energy penalty from the endothermicity of the dehydrogenation reaction. Second, once the hydrogen is released it still needs to be compressed before it can be used to power a truck causing compression losses (~15% energy penalty) and requiring expensive capital equipment.

We address both these problems by developing a powertrain design to support the on-board dehydrogenation of the LOHC using the waste heat in the engine exhaust to drive the endothermic reaction. A computational model is developed that involves four major components: an engine, the dehydrogenation reactor, the waste heat recovery loop/heating systems, and control strategies to meet variable power demand. This model is simulated over the in-house drive-cycle developed for long-haul trucks. Unlike traditional powertrain simulations, our objective is to maximize the sum of the motive force and the waste heat recovered from the exhaust. We consider several design choices to optimize the powertrain design: reactor, heat exchangers, intermediate batteries, turbochargers/superchargers, exhaust after-treatment position, and more. A major focus of this work has been modelling the two-phase flow dehydrogenation reactor necessary for scale-up from bench scale systems reported in literature. Accurate estimates of heat transfer coefficients, and hydrodynamics become particularly important for scale-up where several bench scale assumptions such as uniform bed temperature tend to breakdown. We also perform a detailed study to optimize the engine parameters such as valve timings, spark timings, air-to-fuel ratio, throttle, and more. These variables allow us to control the exhaust temperature and mass flowrate for better heat integration in the overall system. Using the optimized powertrain design, we perform vehicle-level simulations similar to the hydrogen powertrains to obtain TCS and WTW emissions for the LOHC-powered long-haul trucks. Preliminary analyses suggest a trucking system based on this LOHC concept has several practical advantages over competing concepts and would be only modestly more expensive than the current system based on diesel.

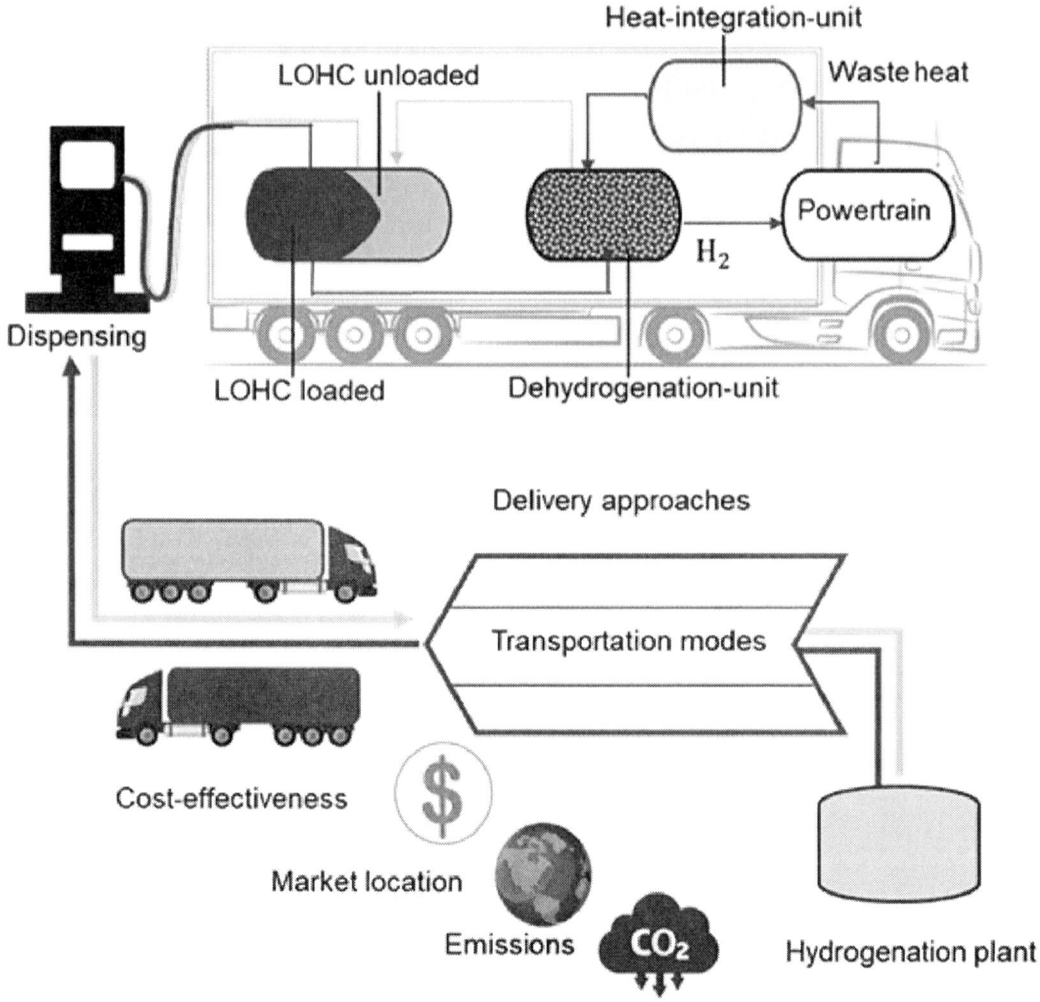

343b Solid-State Hydrolysis of Sodium Borohydride and Sodium Metaborate Hydrate

Geo Jong Kim, Savannah G. Hunt, Allison Millspaugh, Byung Uk Kim, Sang Bum Kim, Hyun-Tae Hwang

Due to the rapid depletion of fossil fuels and pollution from their use, it is becoming more and more important to find an alternative fuel. Hydrogen is considered one of the best contenders as its byproduct is water. To use hydrogen as an effective energy carrier or source, however, safe methods to store hydrogen must be developed. Sodium borohydride ($NaBH_4$, SBH) has attracted great attention as a hydrogen storage material due to its high hydrogen content and inexpensive price. Hydrogen can be generated from the thermolysis or hydrolysis of SBH. Unfortunately, thermolysis requires extremely high temperatures of 300°C or more, while the hydrolysis requires excess amount of water because of low solubilities of SBH and the spent fuel in water and results in safety concern due to the spontaneous generation of hydrogen. In our prior work, it has been demonstrated that thermal hydrolysis of solid-state SBH with boric acid offers improved safety and high hydrogen yield. Despite these favorable outcomes, separation of the products for the regeneration of spent fuel remains a challenge to lower the overall cost. Here, we proposed the hydrolysis of SBH and water formed by thermal dehydration of sodium metaborate tetrahydrate ($NaBO_2$ $4H_2O$, SMBH). The safety hazards due to self-hydrolysis of SBH are eliminated as the mixture of SBH and SMBH is stable at ambient temperature. In addition, SMBH is a product of the hydrolysis of SBH. Since the final products are homogeneous with SMBH, therefore, there is no need for post-reaction separation for the regeneration of the spent fuel. Using this approach, a maximum overall H_2 yields of 4.7 and 5.1wt% was obtained 200 and 250°C, respectively. With high hydrogen yield and simplification of the regeneration process, this proposed method is promising for hydrogen storage for fuel cell applications.

343c End-Use of Hydrogen in Decarbonizing Buildings and Industrial Manufacturing

Praveen Cheekatamarla

Industry and buildings sectors together represent 70% of U.S. primary energy consumption. The generation and use of heat is the most significant end use of energy in both sectors, accounting for 23% of direct GHG emissions. Approximately 90% of manufacturing process heat energy is provided by fossil fuels while 50% of buildings consume natural gas for space and water heating. Decarbonization efforts thus require energy efficient heating and drying technologies capable of operating on low/zero carbon fuels such as hydrogen. Globally, businesses, governments, and academics are working to harness the power of hydrogen. Substitution of hydrogen with natural gas has the potential to eliminate 1500 MMT of CO_2, annually. Novel technologies capable of effectively utilizing hydrogen in heating and power generation applications will be discussed. Technological advancements in direct heating equipment for safely and effectively utilizing hydrogen fuel in both industry and buildings will be presented. The overall value proposition of using hydrogen in combined heat and power applications for deployment in both industries and buildings will be presented.

343d Hydroprocessing of Biomass-Derived Thermal Deoxygenation Oil for Producing Renewable Jet Fuel

Sampath Karunarathne, Sampath Gunukula, Hemant P. Pendse, Clayton Wheeler

Lignocellulosic biomass is an abundant and renewable resource that can be used for producing Sustainable Aviation Fuel (SAF) to reduce greenhouse gas emission in the aviation industry. In this study, lignocellulosic biomass-derived Thermal DeOxygenation (TDO) oil is hydroprocessed and upgraded to jet fuel range hydrocarbons. TDO oil is a type of highly deoxygenated biocrude produced through non-catalytic thermochemical conversion of neutralized biomass acid hydrolysate.

Sawdust from Northern White Pine is converted into a mixture of levulinic and formic acids by acid hydrolysis and dehydration. The organic acids are neutralized with calcium hydroxide to produce organic salts and thermally deoxygenated to bio-crude oil through slow pyrolysis in a pilot-scale continuous flow reactor at 450 °C and atmospheric pressure. The crude TDO oil consists primarily of mono and polyaromatics, and has a broad boiling range (75-550 °C) and a very low oxygen content (< 5 wt %). The TDO oil is hydroprocessed using a commercial Ni/SiO_2-Al_2O_3 catalyst at 300 °C and at 5.2 MPa of hydrogen in a continuous flow tubular reactor. Hydroprocessed TDO oil is distilled into jet fuel range hydrocarbons (180-250 °C), characterized, and compared with commercial and military aviation and turbine fuel standards (ASTM and MIL-DTL standards). The chemical and physicochemical properties of crude TDO oil, hydroprocessed TDO oil, and TDO-jet fuel are analyzed according to the ASTM methods. Further the TDO-jet fuel is blended appropriately into conventional petroleum-derived commercial jet fuel to meet the aviation fuel specifications.

343e Advancing the State of Hydrogen Systems Quantitative Risk Assessment: Design and Requirements of a Hydrogen Component Reliability Database (HyCReD)

Ahmad Al-Douri, Katrina Groth, Kevin Hartmann, Genevieve Saur, William Buttner

Safety and reliability have long been recognized as key issues for the development, commercialization, and implementation of technologies or infrastructure. Reliability engineering and risk assessment play a key role in proactively addressing safety – *before* problems happen – and help us learn from problems *if* they happen. A safe and sustainable transition to expanded hydrogen usage and new applications requires that the risks associated with it be proactively and rigorously investigated, quantified, and mitigated – at the design and early deployment stage when it's cheapest and easiest to engineer-in these features. With hydrogen technologies expected to play a key role in the near future to enable decarbonization of several sectors including energy storage and transportation, the time to address these issues is now. For hydrogen fueling stations, quantitative risk assessment (QRA) is an important tool for enabling their safe deployment and has been increasingly embedded in the permitting process. However, most QRAs to date have used data from non-hydrogen specific data sources. QRAs built on hydrogen-specific component and system reliability data will provide more robust insights, but these data are lacking, thereby hindering the development of necessary safety codes and standards. With 57 operating stations in the United States as well as increasing deployments globally, hydrogen-specific reliability data can be collected and used to update QRA tools.

QRA and reliability data can refer to various types of data depending on the specifics of the analysis being performed. A review of risk and reliability analysis for hydrogen storage and delivery by Moradi and Groth (2019) outlined 8 types of data used in hydrogen QRA [1]. This covers the data needed to perform all steps of a QRA from hazard and exposure identification to frequency and consequence analysis. In another work, West et al.[2] reviewed four existing hydrogen safety data collection tools: H2Tools, Hydrogen Incident and Accident Database (HIAD), and the National Renewable Energy Laboratory's (NREL) Composite Data Products (CDPs). This review found that several elements are systematically missing from current hydrogen data collection practices, including collecting system-level data (component configuration and population), frequencies and component failure data. These elements are necessary to generate failure rates for components of a system, which are a fundamental aspect of QRA and reliability engineering.

To address this gap, a multi-year effort involving a number of research activities is being undertaken at the University of Maryland and NREL with the goal of developing a Hydrogen Component Reliability Database (HyCReD). Based on a review of 10 reliability engineering textbooks and database documentation and guidelines from analogous industries, we developed a set of 23 requirements covering the database characteristics and types of data (static and event) to be collected. The requirements were then used to start an iterative process of identifying and defining the data elements and fields to be part of the database structure. We reviewed the data fields recommended by the references consulted for the developing the requirements and incorporated additional information from two handbooks on reliability data collection in the oil & gas and chemical industries[3], [4]. These references were augmented by monthly discussions with a team of four experts with a total of 70 years of expertise in the field of reliability engineering and hydrogen technologies, and feedback from presentations at academic and industrial conferences as well as multiple internal presentations to stakeholders and potential data providers.

Based on these requirements, we define the structure of the HyCRED database and 25 data elements to be collected, spanning system description, failure, shutdown, or near-miss events, and maintenance events. The data elements are then defined according to international standards used in the safety and reliability practice and potential choice lists are provided for each field. System information refers to the data fields that characterize the facility from which the report originated, its operational environment, and the context of the event being reported. The equipment hierarchy elements are to be completed using drop-down lists from the generic hydrogen station component hierarchy developed by West (2021) [5]. Data elements describing the reported events include date/time, failure mode, mechanism, root cause(s), and severity. Additional event data elements include hydrogen release, release size, accumulation, and ignition are included to support calculating hydrogen-specific leak rates and ignition probabilities in the future.

Since HyCReD is being piloted for hydrogen fueling stations, a generic component hierarchy for these stations based on a representative set of six publicly-available, geographically-varied station designs was developed by West (2021) [5]. This approach combined functional decomposition with service-based decomposition to capture the function of the system as well as all of the necessary components for hydrogen reliability data. Based on the piping and instrumentation diagrams (P&IDs) reviewed, the generic component hierarchy developed detailed 6 subsystems and 21 functional groups with potential components under each functional group. This blended decomposition allows operators, maintenance personnel and reliability engineers to understand component failure in the context of system functions and operating conditions. This is critical to perform process-based risk assessment with a full understanding of system behavior and interactions between system functions, components, and system response as a whole.

Furthermore, a failure mode taxonomy was constructed from an analysis of seven publicly available failure modes and effects analyses (FMEAs) and failure mode sources. While no publicly available FMEAs for hydrogen fueling stations were found, FMEAs for similar systems such as hydrogen fuel cell vehicles and adjacent industries such as oil and gas were consulted. The resulting 33 failure modes and their definitions are initially validated through discussion with a team of five reliability engineering experts and through feedback from presentations as discussed earlier in this section.

To demonstrate the viability of the proposed database, we sought to populate it based on available narrative descriptions of events that occurred in hydrogen fueling stations within the last 10 years. Five events were selected from H2Tools Lessons Learned, the Hydrogen Incident and Accident Database (HIAD), and an incident where we had direct access to the data provider. We determined that almost all of the system data could be readily obtained from the event descriptions. However, it was clear that information on component nominal working pressure, population and time in service (determined from installation date) were not generally reported in these sources. Further, we demonstrated that the component hierarchy was useful in specifying the subsystem, functional group and component that failed for each event. Developing component hierarchies for other hydrogen technologies (e.g., electrolyzers, vehicles, power generation systems) will be key to expanding HyCReD to these systems.

Through the HyCReD structure, we present a major step towards achieving the goal of having hydrogen component and system reliability data to support QRA, safety analysis, reliability analysis, and maintenance planning. The results generated from HyCReD can provide participants and stakeholders with tangible generating insights that enable prioritizing maintenance activities and targeting components with the highest failure probabilities, thereby eliminating costly downtime. This leads to improved station uptime, higher number of fills, and ultimately greater economic returns. Further, these insights lead to improved station designs for the future and enable targeted component research and development activities improving the most failure-prone components. For the wider hydrogen community, HyCReD will generate results that enable risk assessment models that can be used to help establish safety codes and standards. A proactive treatment of safety and reliability can help industry avoid overregulation and facilitate the wider adoption of profitable hydrogen technologies.

343f Ring-Opening of Thermal Deoxygenation (TDO) Oil over Supported Metal Catalysts to Improve Diesel Engine Properties

Oluwaseun Salami, Matthew Kline, Sampath Karunarathne, Clayton Wheeler

While the commercial power sector is improving its carbon footprint, decarbonizing the transport sector still has much room for improvement. Although electric vehicles are becoming more widely used, applications for long distance travel (such as in trucks and airplanes) are still of concern. Renewable crude oil that has been produced by UMaine's Thermal DeOxygenation Process (TDO oil) is an interesting option as a cleaner alternative to conventional diesel and jet fuels. Simple hydrogenation of crude TDO oil is not sufficient because the hydrogenated oil contains many cyclic hydrocarbons, such as cyclo-hexane, decalin and methyl-decalin, that don't perform well in diesel engines. Therefore, the oil must be further upgraded. Breaking the cyclic molecules (ring-opening) can improve the diesel engine combustion properties. In this work, ring-opening is carried out over synthesized catalysts (Ir/SiO_2, Ir/Al_2O_3, Ir/CsBEA and Ir/BEA).

Liters of TDO oil were produced from decomposition of calcium salts of levulinic and formic acids which are made from C_6 sugars in woody biomass by acid hydrolysis and dehydration using dilute sulfuric acid as a catalyst. The TDO oil was then hydrogenated in a downflow reactor packed with Ni/SiO_2-Al_2O_3 catalyst to produce the feedstock for the ring-opening studies. The four novel catalysts which were synthesized in our lab were each evaluated for their ring-opening ability in a downflow, packed bed reactor, maintaining the same catalyst mass, and packing methodology for each experiment. The hydrogenated oil and the ring-opening reaction products were analyzed using a gas chromatograph-mass spectrometer GCMS. The reactor temperature and feed flowrate were varied to determine the optimum conditions to yield a mixture of compounds that are predicted to have the best engine performance characteristics. Gas reaction products were analyzed using a gas chromatograph with a flame ionization detector.

The Ir/SiO_2 catalyst showed the highest conversion to desired ring-opened products such as 1,2-diethylcyclohexane, 1-methyl-2-propylcyclohexane, 1,2,4-trimethylcyclohexane, and least conversion to undesired dehydrogenated and cracked products. This is likely because Ir/SiO_2 is a monometallic catalyst on an inert support. Although side products, such as dehydrogenated products, were formed on this catalyst, conversion to target compounds was still above 50%.

343g NOx-to-Ammonia for Circular Ammonia-Fueled Container Ships

Margarita Charalambous, Alexander Schoch, Gonzalo Guillén-Gosálbez

The shipping industry is highly dependent on fossil-based fuels like heavy fuel oil (HFO), and marine diesel oil (MDO), being responsible for around 2.5% of the global GHG emissions [1]. Notably, decarbonization efforts set by the International Marine Organization anticipate a GHG-reduction of 70% by 2050 [2]. Adopting alternative fuels i.e., low carbon CO_2-based fuels or zero-carbon fuels, is thus imperative. In this context, ammonia, hydrogen, and electricity could be (almost) net zero-carbon fuels that hold a promise to mitigate CO_2 emissions in the shipping industry. However, they face technical challenges, i.e., toxicity, low energy density, and high investment costs [3].

Among the options available, Stolz et al. [4], found that ammonia and methanol are two of the most balanced fuels for the shipping sector in terms of technoeconomic suitability. Recently, MAN Energy Solutions started working in designing a two-stroke ammonia engine, and the auxiliary system that will support the engine. Combusting ammonia requires careful design of emission reduction technologies due to the high expected NO_x emissions, however, there is still no clear view on how NO_x emissions will be handled. Selective catalytic reduction (SCR), has been the main technology that is used to remove NO_x from exhaust gases [5]. Nowadays, ships use SCR to comply with emissions regulations, although the expected large-scale adoption of ammonia-fueled ships might create opportunities for other alternatives to handle NOx emissions.

Motivated by the use of ammonia in two-stroke marine engines, this work applies circular economy principles to reduce NO_x emissions from ammonia-fueled ships and regenerate back ammonia, decreasing the fuel. It was estimated that approximately 34 Mton of ammonia could be produced by using all the NO_x emissions in the exhaust gas from all the power plants in the world. Several studies have looked at the technical feasibility of producing ammonia from NO_x emissions, while still relying (t a lesser extent) on SCR, particularly in power plants [6]. This study will transfer the above idea to the shipping industry. The concept is to regenerate ammonia fuel from NO_x emissions on board an ammonia-fueled ship.

The studied concept involves the modeling of ammonia combustion, the ammonia cracker, NOx-to-ammonia (NTA) reactor, separators, and ammonia tank. The process was modeled in Aspen PLUS v.12. First, a mixture of ammonia/H_2/N_2 generated from ammonia in a cracker is fed to the engine. The ammonia decomposition process is simulated as in Sánchez et al. [7]. The combustion of the mixture has been proposed to reduce the combustion temperature and time, but causes an upsurge of NO_x emissions. Next, the exhaust gases are sent to the NTA reactor where H_2 coming from the cracker is fed to the reactor. The NTA process requires that NO_x reduction with H_2 produces ammonia, although experimental studies have shown that complete selectivity for ammonia is difficult to achieve in stoichiometric conditions. NO_x conversion is enhanced in the presence of CO and H_2O on a Pt/TiO_2 catalyst and at a temperature of 300 ^oC and ambient pressure. The yield of the reaction with CO can be up to 86% [8]. Therefore, we consider that the designed engines operate in dual-fuel mode, and that diesel fuel is mixed at certain operation modes during the ships trip. We further assume that part of the carbon emissions will be used as a source of CO for the NTA reaction [9]. The process simulation of the NTA reactor was based in experimental studies [8]. After the regeneration of ammonia, separation of the gases takes place and ammonia is sent back to the tank.

The production of green ammonia was not simulated, but data for its production were taken from D'Angelo et al. [10].

An economic and environmental assessment was performed as well as a technical feasibility analysis of the developed concept. For the technical feasibility, the weight and space taken by the equipment added on-board the ship is considered and compared to the actual cargo weight (DWT – deadweight tonnage) and space (TEU – twenty-foot equivalent units). In order for the circular concept to be feasible and profitable, the added equipment should not take up most of the space, hindering the ships' purpose. Moreover, the amount of regenerated ammonia dictates the cost of ammonia per tonne·kilometer (tkm).

The environmental assessment was implemented in Brightway2 v.2.4.2, an open-source Python-based LCA software, and its GUI, Activity Browser. The goal of the LCA was to compare the use of ammonia in two-stroke marine engines with and without regeneration of the fuel with the business-as-usual HFO-fueled ships, based on the IPCC 2013 GWP 100a impact assessment method. The scope of the analysis is cradle-to-propulsion, and the functional unit is 1 tkm. The inventories integrate data of the foreground and background systems. For the background system, data available in Ecoinvent v3.8 were used [11]. The mass and energy flows in the foreground system were taken from process simulations, and are complemented with literature data for the production of green ammonia.

Our results show that the circular economy principles could be applied to an ammonia-fueled ship enabling their faster uptake due to, mainly, the possible reductions in investment costs. Our work opens up a new pathway by proposing an alternative concept for tackling the shipping industry's NOx emissions while regenerating the fuel.

Acknowledgments: This work was created as part of the NCCR Catalysis, a National Centre of Competence in Research funded by the Swiss National Science Foundation.

References

1. **IMO *et al.*,** 'Third IMO Greenhouse Gas Study 2014,' *Int. Marit. Organ.*, p. 327, (2014)
2. **IMO *et al.*,** 'Initial IMO GHG Strategy.' (2018)
3. **N. Gray *et al.*,** 'Decarbonising Ships, Planes and Trucks: An Analysis of Suitable Low-Carbon Fuels for the Maritime, Aviation and Haulage Sectors,' *Adv. Appl. Energy*, vol. 1, p. 100008, (2021)
4. **Stolz, B. *et al.*,** "Techno-Economic Analysis of Renewable Fuels for Ships Carrying Bulk Cargo in Europe," *Nat. Energy 2022 72, 7* (2), 203–212, (2022)
5. *"MAN B&W Two-Stroke Engine Operating on Ammonia"* MAN Energy Solutions.
6. **Matsumoto, H. *et al.*,** "Synthesis and Assessment of NOx to Ammonia Conversion Process in Combined Cycle Power Generation Systems," Computer Aided Chemical Engineering, Elsevier, Vol. 49, pp 253–258 (2022)
7. **Sánchez, A. *et al.*,** "Evaluating Ammonia as Green Fuel for Power Generation: A Thermo-Chemical Perspective," Appl. Energy, *293*, 116956 (2021)
8. **Kobayashi, K. *et al.*,** Effect of the TiO_2 Crystal Structure on the Activity of TiO_2 -Supported Platinum Catalysts for Ammonia Synthesis via the NO–CO– H_2O Reaction. *Catal. Sci. Technol.*, *9* (11), 2898–2905 (2019)
9. **Nadimi, E. *et al.*,** "Effects of Using Ammonia as a Primary Fuel on Engine Performance and Emissions in an Ammonia/Biodiesel Dual-Fuel CI Engine" *Int. J. Energy Res.*, *46* (11), 15347–15361 (2022)
10. **D'Angelo, S. C. *et al.*,** "Planetary Boundaries Analysis of Low-Carbon Ammonia Production

Routes," *ACS Sustain. Chem. Eng.*, *9* (29), 9740–9749 (2021)
11. **G. Wernet *et al.*,** "The ecoinvent database version 3 (part I): overview and methodology, *Int. J. Life Cycle Assess.*, vol. 21, no. 9, pp. 1218–1230 (2016)

343h Aspen Plus Simulation for Syngas Conversion with High Pressure Plasma

Dylan Broussard, Varshith Vijayan, Ramalingam Subramaniam

There is little experimental research on chemical processes based on the use of very high pressure plasma. The effects of pressure on the conversion of syngas into hydrocarbons has not yet been reported in any literature pertaining to chemical synthesis accelerated by plasma. While corona and dielectric barrier discharges (DBD), among other plasma technologies, are shown to be more valuable, the utilization of high pressure plasma may represent a novel method to synthesize hydrocarbons in the gaseous phase. The primary goal of this work is to create an ASPEN Plus model of a dielectric barrier discharge system that can forecast performance under a variety of operating situations. Operating parameters vary; variables such as operating pressure and electric current have a considerable big impact. The results reveal that C1-C3 molecules can be synthesized in this manner, and their concentration changes greatly depending on the factors investigated. The plasma-catalysis synergy in this approach shows significant promise for the direct synthesis of value-added compounds and fuels from CO2 and CH4. Additionally, the process's energy efficiency was assessed, and the findings show that the chemical conversion and energy efficiency optimums are congruent.

343i Gaseous Fuels Derived from Landfill-bound Waste- MSW and Plastics- for a Stationary Power Generation Engine: A Performance and Emissions Analysis

Lauren Creadore, Marco J. Castaldi

PROBLEM

In the United States, the predominate end-of-life fate of municipal solid waste (MSW) is landfill with 50% being landfilled, and only 23% being recycled in 2018 (1). Landfilling such volume of waste poses several issues. First, burying the waste precludes its further use as a valuable feedstock for energy production (2, 3, 4), or for valuable extractable materials (5, 6). Second, anerobic decomposition of the buried waste in MSW landfills produces landfill gas (LFG), which, if permitted release to the atmosphere will contribute to the greenhouse gas effect due to the presence of CH4 and CO2.

Municipal solid waste and landfill gas:

Landfill gas is generated in U.S landfills at a rate of 24.2 to 59.1 giga-Nm3 per year (7). LFG uniquely presents both a problem and solution with respect to global warming. Standard practice at landfill facilities is to flare the gas, the products of such combustion having a lower global warming potential than the original gas (8). Without the utilization of the energy released during flaring, there presents a missed opportunity to use the LFG as a working fuel. Direct use of LFG in internal combustion (IC) engines is hindered by low and varying energy density of the fuel. Thus, additions of higher energy density fuels, such as syngas, composed of H2 and CO, improve the flame stability and decrease emissions such as NOX, CO, and unburned hydrocarbons (UHCs) (7).

Plastic waste and pyrolysis gas:

Present in MSW streams are discarded plastics. An estimated 9% of plastics produced are recycled (1). Double digit national recycling rate of plastics has not been achieved as of 2018, leaving the majority of plastic unrecycled as waste. The conversion of end-of-life plastic into liquid fuels via pyrolysis thermal treatment has been a topic of great interest in recent years (9, 10, 11). However, there has been a relative gap in studies on the use of the gaseous products of pyrolysis for use in IC engines for power generation. Co-location of power-generating engines at pyrolysis facilities could aid to offset process energy requirements, thus using waste instead of traditional non-renewable sources for energy.

OBJECTIVES

A variety of gaseous fuel blends having a base of simulated LFG or simulated plastic pyrolysis gas were used to operate a four-stroke spark ignition engine.

With a robust data matrix covering a wide band of realistic operating conditions, including engine load, three objectives were achieved.

1 - Characterization of the fuel blends to quantify emissions.

2 - Identification of lower limit for engine operation (stall condition) for each fuel blend.

3 - Creation of a robust dataset whereby given the system inputs (LFG or pyrolysis gas composition,

maximum expected load), a recommended fuel blend can be referenced for suitable engine operation and desired emission targets.

METHODS

A Honda GC160E 1.6 liter displacement IC engine was retrofitted, with the removal of the carburetor intended for liquid fuel and air mixing, and was operated on a gas either representative of a LFG or a plastic pyrolysis gas.

LFG blends at a range of CO_2/CH_4 ratios representative of field values, which vary naturally due to temperature and waste composition fluctuations (12), were evaluated. Some LFG blends were enhanced with syngas to evaluate the impact on emissions and performance of upgrading the fuel.

Flow rates of the constituent gases for the blends were set using rotameters calibrated with mass flow controllers. Emissions were collected and analyzed with the use of the Enerac 700 emissions analyzer which measures, via a probe which pumps effluent gasses, stack temperature, stack draft pressure, and concentration of O_2, UHCs, CO_2, CO, NOX, NO, NO_2, and SO_2. A Pramac EG2800 electric generator was coupled directly to the engine's power take-off shaft. An electrical circuit comprising of parallel wired light sockets was outfitted with 100W lightbulbs and powered from the generator to apply loading to the engine in increments of 0.2 kW. Engine operation test conditions range from idle (no load) to 2.0kW. A WattsUp Pro99333 power meter was used in-line to verify the power demand.

RESULTS

The discussed experimental setup has been implemented for the LFG-based fuels. Similar experiments on pyrolysis gas fuels are forthcoming.

NOX concentration of the engine exhaust increased with increasing adiabatic flame temperature from 2105K to 2210K. CO and UHC concentrations were found to be coupled. CO and UHCs result from incomplete combustion, which can occur from many causes from poorly mixed air and fuel to lower combustion temperature (7). CO and UHCs increased at engine operation conditions that exhibit poor combustion, such as fuel-starved condition just prior to stalling. The fuel-starved condition resulted in an 878% increase in UHCs compared to the idle condition and was accompanied by degradation in engine operability.

A reduction of the CH_4/CO_2 ratio by reducing CH_4 volume from 84% to 51% greatly increased emissions, with the exception of NO_x. This result is supported by a thermal impact on NO_x, ignition delay impact on UHC and CO, and dilution effect impact on UHC and CO. NO_x emissions decreased by 48% due to a reduced combustion temperature resulting from a large volume of dilutant (49% CO_2). CO emissions increased by 40% when CO_2 composition increased from 16% to 49% and, more significantly, the UHC emissions increased by 2643%.

Tradeoffs between power and emissions were apparent when comparing fuels with an H_2:CO ratio of 0.9:1 and 1:0. Entire CO fraction replacement with H_2 did not appreciably change the power generated. This is due to the similar volumetric energy density between H_2, 12.76 kJ/L, and CO 12.63 kJ/L. However, that replacement resulted in much higher brake specific (BS) total UHCs. The other measured emissions, $BSNO_x$ and BSCO, and the BS fuel consumption were not greatly impacted by replacing the CO fraction with H_2.

Iterations of engine testing were completed with simulated LFG of varying (0-55%) CO_2 fraction, with varying total syngas addition, and with varying H_2 fraction in the syngas. Data was compared across test cases to identify engine operational limits and regions of performance interest.

IMPLICATIONS

When engaging the waste management hierarchy (1), a move from the least preferred method of treatment and disposal (landfill) to a more preferred method of energy recovery, can be possible with direct utilization of either LFG or plastic pyrolysis gas. Compared to status-quo alternatives of flaring LFG (which captures no benefit from its inherent energy content) and of landfilling hard-to-recycle plastics, energy generation offers a more intentional resource use pathway.

Utilization of LFG-based fuels could be of interest especially in countries with high landfilling rates, including the United States. Utilization of plastic pyrolysis gas-based fuels should be investigated as the number of plastic pyrolysis facilities continues to grow (13) to more fully understand the environmental implications. Should such fuels be determined viable for direct use as drop-in power sources, costly changes to existing equipment and infrastructure could be largely avoided.

More widely, the methods outlined could be extended to upgrade other fuels beyond those discussed here.

REFERENCES

1. Olem, National Overview: Facts and Figures on Materials, Wastes and Recycling. US EPA, (2022).

2. About Partners of the Landfill Methane Outreach Program. US EPA, (2022).

3. H. R. Amini, D. R. Reinhart, Regional prediction of long-term landfill gas to energy potential. Waste Management 31, 2020-2026 (2011).

4. L. Levaggi, R. Levaggi, C. Marchiori, C. Trecroci, Waste-to-Energy in the EU: The Effects of Plant Ownership, Waste Mobility, and Decentralization on Environmental Outcomes and Welfare. Sustainability 12, 5743 (2020).

5. M. J. Castaldi, J. LeBlanc, A. Licata, The Case for Waste to Energy. Mechanical Engineering 144, 34-39 (2022).

6. S. C. Gutiérrez-Gutiérrez, F. Coulon, Y. Jiang, S. Wagland, Rare earth elements and critical metal content of extracted landfilled material and potential recovery opportunities. Waste Management 42, 128-136 (2015).

7. M. P. Kohn, M. J. Castaldi, R. J. Farrauto, Auto-thermal and dry reforming of landfill gas over a Rh/γAl2O3 monolith catalyst. Applied Catalysis B: Environmental 94, 125-133 (2010).

8. O. Ayalon, Y. Avnimelech, M. Shechter, Solid Waste Treatment as a High-Priority and Low- Cost Alternative for Greenhouse Gas Mitigation. Environmental Management 27, 697-704 (2001).

9. I. Kalargaris, G. Tian, S. Gu, The utilisation of oils produced from plastic waste at different pyrolysis temperatures in a DI diesel engine. Energy 131, 179-185 (2017).

10. K. Kumar, R. Puli, Effects of Waste Plastic Oil Blends on a Multi Cylinder Spark Ignition Engine. MATEC Web of Conferences 108, 08005 (2017).

11. Khairil et al., Experimental Study on the Performance of an SI Engine Fueled by Waste Plastic

Pyrolysis Oil–Gasoline Blends. Energies 13, 4196 (2020).

12. L. Hailing et al., Seasonal CH_4 and CO_2 effluxes in a final covered landfill site in Beijing, China. Science of The Total Environment 725, 138355 (2020).

13. S. Hann, T. Connock, 'Chemical Recycling: State of Play,' (Eunomia, 2020).

343j Influence of CO_2 Concentrations on *Chlorella Kessleri* Cultivation, Growth, and CO_2 Fixation

Shaikh Razzak

CO_2 biofixation by microalgae has garnered considerable interest. In this context, culture of *Chlorella kessleri* one of the microalgae species in batch photobioreactors with varying levels of CO_2 in the air. This study aims to determine *Chlorella kessleri*'s capacity for CO_2 biofixation and biomass growth. The biomass concentration and biomass productivity of *Chlorella kessleri* cultivated under various CO_2 concentrations at 25^0C temperature and nitrogen to phosphorus (NP) ratio of 2:1 were investigated. During the culture period, biomass concentration increased progressively for all concentrations of 2%, 6%, and 10% CO_2 examined, whereas the opposite is true for 0% CO_2 concentration. Maximum biomass concentration and productivity were determined to be 607 mg/L and 46 mg/L/d, respectively, at 2% CO_2 concentration in the air. Biofixation of CO_2 is primarily dependent on the species and optimal CO_2 concentration provided to the culture. Maximum CO_2 biofixation rate of 84 mg/L/d was achieved at 2% CO_2 air concentration. In conclusion, the culture of *Chlorella kessleri* microalgae in photobioreactor is a potential strategy for the simultaneous removal of nutrients and biofixation of carbon dioxide.

369b Effect of Gas Composition on Co-Electrolysis Performance of Solid Oxide Electrolyzer Cells

Masashi Kishimoto, Riki Nakamura, Hiroshi Iwai

Solid oxide electrolysis cells (SOECs) are attractive electrochemical energy conversion devices with high conversion efficiencies because of their high operating temperatures of around 700°C. They can convert electrical energy into chemical energy by electrochemically reducing steam and carbon dioxide to produce hydrogen and carbon monoxide, which enables large-scale energy storage. The mixture of hydrogen and carbon monoxide is also an important precursor to produce hydrocarbon fuels through methanation and the Fischer-Tropsch processes. Since the electrolysis performance of the SOECs depends on operation conditions such as temperature and gas compositions, their effects have been investigated experimentally in the literature. In particular, the electromotive force (EMF) of the cells and the electrochemical activity of the fuel electrode are essential to understand the cell performance. They are also desired to be formulated into mathematical models to be used for the numerical simulation of SOECs.

The EMF is usually estimated using the Nernst equation based on the gas compositions on the electrodes, namely, the anode and cathode. However, unlike the steam electrolysis operation, several redox couples co-exist in the cells under co-electrolysis operation, inhibiting the simple use of the Nernst equation. When steam and carbon dioxide are supplied to the fuel electrode of SOECs, the reverse-gas-shift reaction produces a gas mixture consisting of hydrogen, steam, carbon monoxide, and carbon dioxide, and its composition is not always at equilibrium. Therefore, evaluating the EMF is not a trivial problem and requires experimental and theoretical investigation.

The electrochemical reaction in the fuel electrode is also considered to depend on gas compositions. It is reported that carbon dioxide is less actively reduced in the co-electrolysis operation; hence steam is preferentially reduced. In this case, the reverse-gas-shift reaction is the main route for converting carbon dioxide to carbon monoxide. However, comprehensive analyses have not been conducted to understand the dependence of the electrochemical activity of the fuel electrode on gas compositions.

Therefore, in this study, the EMF and the electrochemical activity of the fuel electrode are experimentally investigated under various gas compositions with a particular focus on the effect of the reverse-gas-shift reaction. A catalyst pellet is inserted near the fuel electrode to offer a larger reaction site and control the progress of the reverse-gas-shift reaction in the supplied gas. The EMF, current-voltage, and impedance characteristics are measured under various gas compositions.

The electrolyte-supported SOEC cells are fabricated as follows and used in this study. A yttria-stabilized zirconia disk (8-YSZ, Tosoh Corp., Japan) with a diameter of 30 mm and a thickness of 500 μm is used for the electrolyte. Ni-YSZ anode is fabricated using screen printing. NiO powder (Fujifilm Wako Pure Chemicals Corp., Japan) and YSZ powder (TZ-8Y, Tosoh Corp., Japan) are mixed by planetary ball milling with ethanol. The mixture ratio is determined to achieve Ni: YSZ = 50: 50 vol.%. After evaporating ethanol, the mixture powder is sieved to remove agglomerate and mixed with a binder (VEH, Nexceris LLC, USA) at NiO-YSZ: binder = 75: 25 wt.% to form anode slurry. The obtained slurry is screen printed on the electrolyte surface using a mask with a thickness of 60 μm and sintered at 1400°C for 5h. On the other side of the anode, a barrier layer consisting of GDC (gadolinia-doped ceria, GDC-10, Shin-Etsu Astech Co. Ltd., Japan) is formed by spin coating. GDC powder is mixed with binder at GDC: binder = 25: 75 wt.% and dropped onto the YSZ disk rotating at 8000rpm. Sintering is performed at 1300°C for 10h. On the GDC barrier layer, LSCF (LSCF-6428-N, Kceracell Co. Ltd., Korea)-GDC cathode is screen printed using a mask of 60 μm and sintered at 950°C for 5h. In addition, to increase the catalytic activity of the fuel electrode for the reverse-gas-shift reaction, a Ni-YSZ pellet is inserted near the fuel electrode to promote the reverse-gas-shift reaction. The pellet is fabricated using a tape-casting process. Details of the fabrication process are found elsewhere (Seo et al., Fuel Cells, 2020).

The electrochemical performance of the cell is evaluated at 800°C with several gas compositions. For the co-electrolysis experiment, H_2: H_2O: CO_2: N_2 = 10: x: 40-x: 50 (x = 0, 10, 20, 30, 40) on the fuel electrode, and pure oxygen on the oxygen electrode is used. Note that the total amount of steam and carbon dioxide is kept constant. Hydrogen is added to avoid the oxidation of nickel. For the reference experiment, steam electrolysis and carbon oxide electrolysis are also performed under H_2: H_2O: N_2 = 10: x: 90-x (x=10, 20, 30, 40), and H_2: CO_2: N_2 = 10: x: 90-x (x=10, 20, 30, 40), respectively. The open-circuit voltage (OCV) and current-voltage and impedance characteristics are measured using a Solartron 1455A frequency response analyzer and a Solartron 1470E electrochemical interface (Solartron Analytical, UK). The outlet gas compositions are measured using a gas chromatograph (990 Micro GC System, Agilent, USA).

First, OCV is measured with and without inserting the Ni-YSZ catalyst near the fuel electrode (Fig. 1). Overall, OCV increases as the amount of carbon dioxide increases in the supplied gas. The OCV obtained from the cell without the catalyst is found to be closer to the EMF obtained using the Nernst equation considering the hydrogen and steam redox couple. On the other hand, the OCV obtained from the cell with the catalyst is lower than in the other case and closer to the EMF obtained using the Nernst equation with chemical equilibrium of the reverse-gas-shift reaction. These indicate that the reverse-gas-shift reaction is not always at the chemical equilibrium in the gas phase and that the hydrogen-steam redox couple has a more significant influence on the OCV. Subsequently, the mixed potential theory is employed to develop a model for the EMF. The theoretical EMF values obtained using the Nernst equation from the hydrogen-steam redox couple and the carbon monoxide-carbon dioxide redox couple are weighted averaged. The surface coverages of the gas species are calculated from the adsorption equilibrium and used for the weights. By calibrating the equilibrium constants within a physicochemically reasonable range, the developed model successfully reproduces the trend of EMF in the experiment. In addition, the obtained surface coverage of the gas species indicates that the electrochemical reaction site is mainly covered with hydrogen and steam, particularly when the catalyst is not used, which is why the OCV is closer to the value obtained from the hydrogen-steam redox couple.

Next, the current-voltage and impedance characteristics are measured under co-electrolysis operation and compared with those under steam electrolysis and carbon dioxide electrolysis. The overpotential under carbon dioxide electrolysis is found to be larger than that under steam electrolysis. The overpotentials under co-electrolysis operation are found to fall between those of the steam and carbon dioxide electrolysis but are closer to those under steam electrolysis. These indicate that the carbon dioxide is less electrochemically active in the electrolysis. A similar trend is observed in the impedance measurement. The polarization resistances obtained from the equivalent circuit fitting (Fig. 2) are larger in the carbon dioxide electrolysis compared with the steam electrolysis, and the values in the co-electrolysis are closer to those in the steam electrolysis.

In summary, during the co-electrolysis operation of SOECs, it is found that the EMF and polarization resistance are mainly determined from the hydrogen-steam redox couple because of the higher surface coverage of hydrogen and steam near the reaction site.

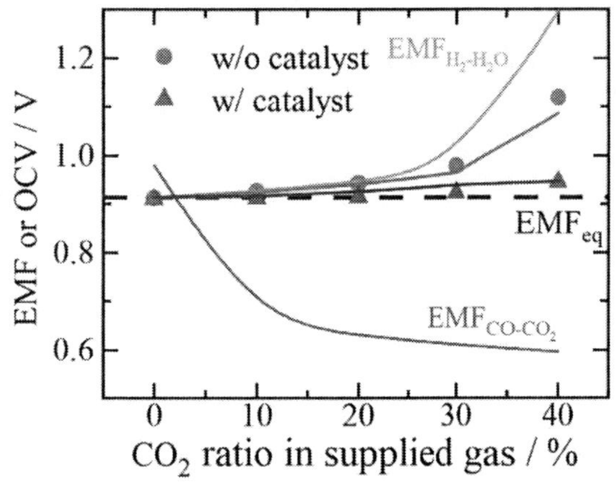

Fig. 1 Open-circuit voltage in eqperiment and electromotive force based on Nernst equation and Mixed potential theory.

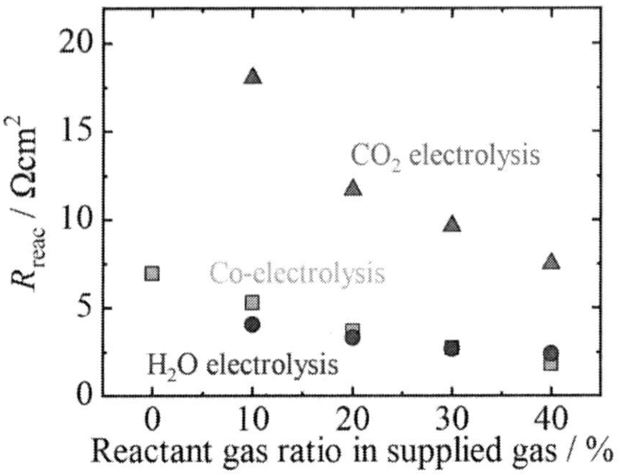

Fig. 2 Polarization resistance obtained from equivalent circuit fitting analysis.

369c Advances in High Temperature Electrolysis for Sustainable Hydrogen Production: Insights from Idaho National Laboratory's Renewable Energy Research Portfolio

Joshua Gomez

Solid oxide electrolysis cells (SOECs) are high-temperature ceramic devices that can convert electrical energy into chemical energy by splitting water molecules, offering a more efficient way of producing hydrogen than conventional methods. The high operating temperatures of SOECs offer greater thermodynamic efficiency at larger scales compared to other hydrogen production technologies, making them an attractive option for transitioning towards a hydrogen economy, as many governments worldwide have expressed great interest in recently. Idaho National Laboratory (INL) has steadily increased its research and testing efforts toward SOECs over the past decade, making it a leading institution in developing and advancing hydrogen and electrolysis technologies. INL's hydrogen research portfolio encompasses a range of capabilities, including high-throughput fundamental novel material/manufacturing research, accelerated stress testing of button cells, and the capacity to pilot test large-scale 25-250 kW SOEC systems. Fundamental materials research is essential for identifying novel materials with enhanced SOEC performance, such as improved conductivity and durability, or novel electrolytes, such as proton-conducting materials. INL applies advanced techniques to model and simulates SOEC unit operations, helping to identify optimal process parameters and design features. Accelerated stress testing (AST) of button cells and large planer cells used stacks allows for evaluating cell performance under real-world operating conditions. AST can be used to determine the effects of various stressors, including temperature, humidity, and modulated electrochemical testing. This information can then be used to optimize SOEC performance, identify key degradation factors, and recognize potential design or operational issues before full-scale implementation. Meanwhile, the pilot testing of large-scale SOEC systems provides valuable insight into the feasibility of commercial-scale production of novel stack designs. In addition to these core research areas, INL's hydrogen research portfolio includes investigations into advanced catalysts and reactor design, process optimization, control, and integrating SOECs with renewable/nuclear energy sources. State-of-the-art process engineering and quality assurance/quality control (QA/QC) measures ensure that findings are scientifically robust and practical, making them relevant to real-world applications. Overall, INL's hydrogen research portfolio is diverse and comprehensive, encompassing various capabilities and research areas.

369d Development of a New Packed-Bed Type Electrolyzer for the Production of CO or Syn-Gas through Electrochemical CO_2 Reduction Reaction

Donggu Han, Ung Lee

Electrochemical CO_2 reduction (eCO_2R) is a promising alternative method for producing CO or syn-gas in a sustainable manner. Nonetheless, numerous obstacles must be surmounted prior to the realization of commercial viability. Many studies focused on the direct conversion of CO_2-captured amine solutions, which utilize amine absorbents to capture CO_2 from flue gas emissions, into value-added compounds are receiving increased interest because considerable thermal energy must be provided to regenerate the absorbed CO_2 to obtain pure CO_2. Here, we present a packed-bed type electrolyzer designed to allow CO_2-captured amine solutions, as a catholyte, to pass through a flow path packed with three-dimensional electrodes made of carbon felt-binder-silver (Cfelt-b-Ag). For these three-dimensional electrodes, the local hydrophobicity inside the carbon felt can be controlled by adjusting the type and amount of binders. A well-controlled hydrophobicity of the three-dimensional electrodes can facilitate the formation of an instantaneous three-phase interface around the silver surface, which promotes eCO_2R. In a system with counter anodes of only 10 cm^2 scale, we achieved a total current of 15 A while maintaining over 50% faradaic efficiency of the CO.

369e Assessing and Improving Electrode Structures for Scalable CO_2 Reduction Using *in Operando* electrochemical Techniques

Tim Van Cleve, Prantik Saha, Danielle Henckel, Leiming Hu, K.C. Neyerlin

Electrochemical CO_2 reduction is a promising technology to capture and convert CO_2 to valuable chemicals. High Faradaic efficiencies of CO_2 reduction products can be achieved with zero-gap alkaline CO_2 electrolyzers with a supporting electrolyte at the anode (anolyte) at current densities above 250 mA/cm^2 [1]. While much of the field has focused on tailoring the electrocatalyst structure and electrochemical interfaces to minimize kinetic overpotentials, there is a broader need to understand how different cell configurations can impact the observable electrode properties and performance. For example, we have found the presence of KOH at a copper-coated gas diffusion electrode significantly enhances its capacitance (proxy for electrochemically active surface area) and ionic conductivity by 4 and 447 times, respectively [2]. Additionally, the low coverage of Nafion in these electrodes suggests the ionomer has a very limited role in tuning the electrochemical interface in these electrodes [2].

This presentation will describe our methodology to study CO2 electrodes in operando independent of the material selection and cell configuration. We use electrochemical impedance spectroscopy to monitor changes in electrode properties at different relative humidities and in different cell configurations to elucidate key performance-limiting phenomena.

References

[1] Y. Chen, et al. ACS Energy Letters 2020 5 (6), 1825-1833.

[2] P. Saha et al 2023 J. Electrochem. Soc. 170 014505

449a Introduction to Natural Gas Production and Processing

M. R. Riazi

World production of natural gas in 2021 was more than 144 trillion cubic foot a 4% increase from 2020 production and according to IEA consumption of natural gas is expected to grow by 1% annually. In addition to direct use as a fuel, natural gas is converted into LPG, many petrochemicals and hydrogen. In 2019 before the pandemic, more than 23% of world energy was produced by natural gas after oil and coal. With current production and proven reserves, it is expected to last for 52 years excluding unproven reserves. It is the cleanest source of fossil type energy and currently provides the most economical and commercial method of hydrogen production by steam reforming. Today's technology choices and the upgrading of petrochemical facilities, however, need to be driven by a completely different logic than the existing industry. Current operations were built on the basis that economics of carbon dioxide are only due to fuel cost, however, the future of regulations and market acceptance look very different in the future in many countries, and these changes are driving big financial decisions within the industry. Recent advances in pyrolysis of natural gas provides a promising commercial method for production of hydrogen while eliminating GHG emission.

449b Gas Hydrate Based Solidified Natural Gas (SNG) Technology Incorporating Dual-Function Promoters for Enhanced Commercialization Potential

Gaurav Bhattacharjee, Ye Zhang, Praveen Linga

Natural Gas (NG) has gained newfound importance in recent years as a crucial transition fuel, also spurring the inclusion of NG storage techniques within nations' energy resilience frameworks. The gas hydrate based Solidified Natural Gas (SNG) technology is an attractive option for large-scale, long-term NG storage applications due to its ability to provide non-explosive, energy-dense storage with straightforward gas recovery. Although gas hydrate formation generally necessitates low temperature and high pressure conditions, utilizing dual-function promoters (additives that promote both the thermodynamics and kinetics of hydrate formation), rapid and energy-dense methane hydrate formation may be achieved at ambient temperature.

We have identified three different additives, namely Tetrahydrofuran (THF), 1,3-Dioxolane (DIOX) and 1,3-Dioxane (Dioxane), as dual-function promoters for the methane hydrate formation process, and thus, the SNG technology. Herein, we collate data gathered over several years of systematic research to present a holistic comparison of the performances and efficacies of these dual-function promoters. Our reportage first compares the critical physico-chemical properties of the three compounds and then focuses on their 'dual-impact' on the thermodynamics and kinetics of the mixed hydrate systems. For the latter aspect, phase-equilibrium conditions for the mixed methane hydrate systems are compared and reported, as is relevant kinetic data such as the time required to complete 90% of hydrate formation (t_{90}) and the final gas uptake. Finally, the stability characteristics of individual mixed methane hydrate pellets, synthesized using the different dual-function promoters, are discussed. The present study suggests a new direction to enhance the operational and commercial feasibility of the SNG technology.

449c Pore-Scale Methane Hydrate Depressurization in Confined Pores for Efficient Methane Recovery

Qian Ouyang, Jyoti Shanker Pandey, Nicolas von Solms

Natural gas is a cleanest-burning and energy resource with high energy density. Natural gas hydrate (NGH) is a crystalline compound formed by water molecules and natural gas molecules under low temperature and high pressure. It is estimated that 10^{14}-10^{18} m^3 natural gas (main component methane) is stored in NGH reservoirs, most of which are located 0-110m below seafloor. Massive oceanic methane reserves in NGH have attracted worldwide attention for utilizing this prospective future resource. It is therefore of importance to develop efficient, safe and economical exploitation method to recover methane from hydrate reservoir.

Depressurization, as the most feasible hydrate exploitation method, has been employed in many field tests. However, several technical issues need to be harnessed to facilitate its efficiency for commercial exploitation in the near future. One of the key issue is hydrate reformation. Hydrate dissociation is an endothermic process and depressurization may induce reservoir temperature to reduce. Loss of heat supply in hydrate-bearing sediments causes methane hydrate reformation near wellbore, which can significantly affect fluid mobilization, plug the pores in sediments and reduce methane gas productivity. Therefore, it is necessary to study the distribution habits and morphological features of secondary methane hydrate in sediments, with the purpose of avoiding or tackling hydrate reformation along with hydrate exploitation.

This work employed a micromodel etched with pores to investigate methane gas recovery from hydrate-bearing sandstone. The dynamic process of hydrate transitions and gas/water migration can be visualized directly in fixed field of view. Secondary CH_4 hydrate formation was initialized after multiple cooling & heating and instant gas/water production at constant-pressure. Depressurization exploitation on methane hydrate was conducted by constant-rate pressure depletion on micromodel. The focuses of this work were to: (1) confirm whether secondary CH_4 gas hydrate formation happened during gas/water production in hydrate-bearing sandstone; (2) how gas/water/hydrate varied in morphology during secondary formation and following constant-rate depressurization; (3) illustrate the effects of seepage properties, such as wettability and fluid saturation, on secondary hydrate formation and dissociation that affecting methane gas recovery efficiency.

The results showed that hydrate nucleation happened at gas/water interface simultaneously or 5-20 minutes after gas/water flow at 80 bar and 1°C. The hydrate nuclei then grew towards gas phase by consuming gas available, with the hydrate patterns varying from coarse to smooth totally in hydrophilic gas-rich pores and hydrophobic water-rich pores. During constant-rate depressurization on secondary methane hydrates, pore blockage with pressure differences of 3-45 bar existed until pressure relief existed at dissociation pressures of 28-45 bar. Abnormal methane hydrate dissociation and reformation emerged because of localized pressure variation triggered by gas/water migration in confined pores. These kinetic and morphology findings are beneficial to understand the mechanisms of hydrate transitions and fluid migration in confined pores, and thus provide insights of reducing secondary methane hydrate formation for efficient methane gas recovery through controlled depressurization.

449d Determination of Local Crystal Orientation of Methane Hydrates in Tailored Confined Media Using Polarized Raman Spectroscopy

Mrityunjay Sharma, Wiwin Febrianto, Yuan Yang, Carolyn Koh, Ryan Hartman

Naturally occurring gas hydrate molecules are energy-rich entities where guest molecules are enclosed in polyhedral host cages of hydrogen-bonded water molecules. Methane hydrates form at high pressure and low temperature. The formation of hydrates is a nucleation and growth phenomenon where molecular-level understanding of chemical physics is of principle importance. Though, in nature, gas hydrates crystallize in the confined media, most reports are available only for the bulk system. Also, the influence of confinement on cage occupancy and different hydrate structure is not well understood. Along these lines, a microfluidic reactor was fabricated with a nano-porous structure placed in the device's channels. The Methane hydrates were crystallized in the nanopores, which act as the confined media, and the orientation of crystals was analyzed using polarized Raman spectroscopy. Using polarized Raman spectroscopy, a detailed methodology is proposed for determining the crystal orientation of the Methane hydrates in the nanopores. Three different diameters (10, 20, and 40 nm) of nanopores were used to analyze the confinement effect on the crystal's orientation and cage occupancy. The impact on crystallization in bulk and nanopores was compared based on the stability of crystals.

449e Commercial Applications of Petroleum Data Analytics Using Shale Oil and Gas Data Sets from Permian Operator

Vyacheslav Romanov, Guoxiang Liu, Shahab D. Mohaghegh, Xiongjun Wu, Derek M. Vikara, Kolawole Bello

This work was focused on developing novel methods to facilitate rapid data to knowledge transformation for subsurface applications, particularly those related to unconventional oil and gas. It was conducted at the U.S. Department of Energy's (DOE) National Energy Technology Laboratory (NETL) in collaboration with industrial partner, a major shale-field operator in the Permian Basin, located in West Texas, United States. With progressing oil and gas field maturation and infill drilling, tighter well spacing to maximize resource utilization may lead to inter-well interference, to the point that induced fracture hits may start to affect the efficiency of fluid recovery. Fracture hit (aka, 'frac hit') is the form of hydraulic inter-well communication where hydrocarbons production at an existing well is affected by pumping hydraulic fracturing fluids into another well. Petroleum data analytics was used in this work to demonstrate the opportunities to optimize field development planning, including the frac-hit detection and analysis, support operational decision-making and improve resource recovery. It involved descriptive and predictive analytics which generated fuzzy patterns and key performance indicators for site characterization, completion design, hydrocarbons production, and frac-hit model implementation. Random Forest (RF) was used as the base estimator in the feature importance assessment, to evaluate the model input contributions to production response and to appraise data value. The demonstration also involved a combination of several Machine Learning (ML) algorithms such as Fuzzy Set Theory and Artificial Neural Networks (ANN). The ANN models included long short-term memory (LSTM), recurrent neural networks (RNN), and multilayer perceptron (MLP) neural networks to innovatively identify hydraulic communications between fracking child well(s) and producing parent well(s) in the same pad (intra-pad interactions) and/or in different pads (inter-pad interactions) based on pressure and production measurements in specified time series. The models took advantage of LSTM in handing temporal data and the MLP in feature extraction. The developed workflow can capture the time-variable features of frac hits when the model is deep and wide enough with enough trainable parameters for learning and feature extraction. The number of model parameters ranged from approximately 44,000 to 916,000. The five-fold increase in the number of model parameters (from the smallest to mid-size model) resulted in almost three orders of magnitude decrease in the mean-squared-error (MSE). The additional four-fold increase in the number of model parameters (from mid-size to the largest model) resulted in another order of magnitude decrease in MSE. As expected in time-series applications, temporal attributes such as the current-day lift-inlet pressure, daily lag pressure, and days online had orders of magnitude greater importance than static and categorical attributes. The results also demonstrated that trained ML models can accurately predict the frac-hit probability in near-real-time, with major implications for a potentially real-time operational support. The observed patterns and correlations provide a good starting point for determining the likely cause of observed shut-in an frac-hit events.

449f Tuning Activity and Selectivity in the Partial Oxidation of Methane over Iridium and Ruthenium Mixed Oxides Supported on Titanium Dioxide Nanoparticles

Helena Hagelin Weaver, Li-Yin Hsiao

Selective oxidation of natural gas to higher value chemicals in a one-step reaction process would be highly desirable in a transition away from petroleum-based chemicals and fuels. However, avoiding overoxidation to the most thermodynamically stable oxidation products, i.e. carbon dioxide and water, is challenging, partially due to the high temperatures needed to break the strong C-H bond of methane, the major constituent of natural gas. Iridium dioxide is a very promising material that, under certain conditions, can activate methane at room temperature. However, at room temperature catalytic turnover is challenging, and higher temperatures are needed for the catalytic formation of oxidized products, temperatures at which overoxidation tends to prevail. One strategy to avoid overoxidation is to add a less active (more selective) oxide to an unselective and highly active oxide catalyst. We have therefore added RuO_2 to IrO_2 in an attempt to better control the oxidation properties of the resulting catalysts. Due to the cost of these active metals (Ru and Ir), low loadings (0.5% by weight) were utilized, and the ruthenium and iridium oxides were supported on commercial rutile nanoparticles. The temperature programmed reduction (TPR) data (using 5% H_2 in N_2) reveal that the temperature of reduction is indeed different and varies with the Ir:Ru ratio (Figure 1). Preliminary reaction data run under stoichiometric conditions for the partial oxidation of methane to synthesis gas (carbon monoxide and hydrogen), i.e. CH_4/O_2/Ar/He at 2.0/1.0/2.0/95 standard cm^3/min and 50 mg of catalyst, indicate that both the activity (CH_4 conversion) and selectivity to CO and H_2 in this probe reaction are influenced by the Ir:Ru ratio.

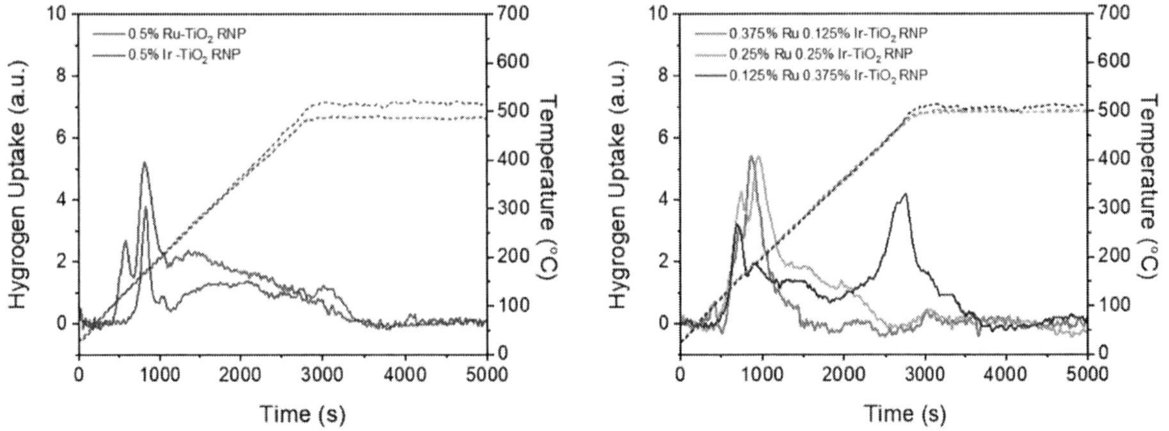

Figure 1. Temperature programmed reduction data collected from Ir and Ru catalysts supported on rutile TiO_2 nanoparticles (RNP) as a function of Ir:Ru ratio.

551a Heavy Oil Reforming in a Dual Circulating Fluidized Bed Reactor

Girish Srinivas, Steve Schwab

TDA Research Inc. (TDA) has developed, patented, and demonstrated a chemical looping steam reforming process (which we call *HyRes* – "Hydrogen from Residuum") that can convert liquid hydrocarbon fuels into syngas from which hydrogen can be extracted. *HyRes* can generate syngas from a wide variety of feedstocks including JP-8, gasoline, kerosene, propane, natural gas, heavy oils (including vacuum residuum) and even tar sand bitumen. While it is generally considered impossible to steam reform vacuum residuum without catalyst deactivation, the reason that our process works is because the catalyst is continually regenerated in air to burn off deposited sulfur and coke before too much can build up. All liquid hydrocarbon fuel based reforming operations are prone to severe catalyst coking when using fixed bed steam reforming. The major advantages of our process over fixed bed reforming include: the ability to use waste/dirty fuel sources because of our feed-agnostic reactor design, the catalyst does not degrade because it is continuously regenerated before contaminants build and deactivate it, and the majority of the sulfur in the feed is removed from the syngas product.

When our *HyRes* process is fully integrated with trace sulfur removal, water gas shift (WGS), and pressure swing absorption (PSA) units, the result is an efficient process to generate purified hydrogen that can utilize a wide range of fuel feedstocks and be scaled for a wide range of applications. TDA is currently contracted by the Department of Defense (Contract # N68335-23-C-0093) to scale-up our *HyRes* technology to produce 10 kg/day of H_2 on a skid-mounted unit capable of being transported between operational sites to generate purified H2 for fuel-cell applications. In this presentation at the 2023 AIChE Annual Meeting, we will discuss the ongoing pilot-scale skid-mounted reformer system, present additional operating data on this technology, and discuss the market opportunities for the technology.

In TDA's *HyRes* process, hydrocarbon feedstock and steam are fed to a fluidized bed of Ni-based steam reforming catalyst at 850°C. The hydrocarbons are completely converted into an equilibrium composition product syngas stream. The catalyst is continually transferred to the regenerator, where it is fluidized with air to burn off any coke while the nickel is converted into nickel oxide. In the reforming step, the sulfur in the fuel is removed by the nickel as nickel sulfide (which is burned off in the regenerator and does not enter the hydrogen stream). During regeneration, the catalyst is heated by the oxidation of Ni to NiO. Both Ni oxidation and extra fuel combustion in the regenerator heat the catalyst so that the sensible heat that the catalyst picks up during regeneration supplies energy to the endothermic steam reforming reaction taking place in the reformer. Our catalyst does not deactivate because regeneration removes coke and sulfur from catalyst before they can reach damaging concentrations. Upon returning to the reformer the NiO is reduced back into catalytically active Ni metal by the hydrocarbons in the feed, and steam reforming continues.

TDA's process for H_2 production from fuels uses a commercially available Ni based steam reforming catalyst, but the key to our process is the system design, not the catalyst. The process is based on a dual circulating fluidized bed (DCFB) reactor system, which lets us process dirty, highly coking feedstocks at low steam-to-carbon ratios (low steam-to-carbon ratios improve the energy efficiency of the process). The DCFB reactor design allows for catalyst to continuously flow around the reactor loop while sealing off the gas environments in the reformer and regenerator from each other. By utilizing fluidized bed steam reformers, the sulfur and coke that deposit on the catalyst during reforming forms only a thin film on the surface of the catalyst particles, in contrast to the thick crust of coke that would be formed in a fixed bed, making regeneration easier.

Critically, the sulfur compounds in the feed react with the nickel catalyst in the reformer bed, forming nickel sulfides. Because the sulfur is chemically bonded to the catalyst, the product syngas contains minimal sulfur, which helps keep sulfur out of the downstream H_2 purification processes. While the sulfur temporally poisons the catalyst, it is quickly transferred to the regenerator before it becomes irreversibly fouled (the reformer residence time is ~7 minutes, far shorter than the time it would take to poison all the catalyst with sulfur or foul it with coke). We have successfully processed very high-sulfur (up to 3.4 wt.%)

heavy feedstocks such as vacuum residuum and HVGO without catalyst deactivation in our DOE Phase II projects. Therefore, in TDA's *HyRes* process, the Ni catalysts simultaneously acts as a steam reforming catalyst and sulfur sorbent. As a result, only a small amount of sulfur (<8 ppm H2S) needs to be further removed from the product syngas (using a disposable sorbent) to obtain completely sulfur-free syngas.

The first reactor is the steam reformer (Figure 1, right), where fuel reacts with steam at approximately 850°C to produce syngas (to be subsequently shifted and purified into H_2). The heat needed to drive these reactions is provided by the flow of hot catalyst particles from the regenerator. As regenerated catalyst flows into the reformer reactor, used catalyst flows out and is transferred to the regenerator by gravity. A loop seal, which is a nonmechanical valve that transfers solid catalyst particles using aeration gas, facilitates the catalyst flow to the regenerator. The second reactor vessel is a catalyst regenerator (Figure 1, bottom left) where coke and sulfur are burned off in air, which partially reheats the catalyst. The Ni metal in the catalyst is also oxidized to NiO during this step, and this provides the majority of the energy which heats the catalyst. The hot catalyst travels up the riser, is separated by the cyclone, and flows back to the steam reformer where it provides the endothermic heat of reaction. In the reformer, the NiO is reduced back to catalytically active Ni metal by hydrocarbons and syngas, and steam reforming of the fuel resumes. Thus, the sensible heat of the hot catalyst returning from the regenerator to the steam reformer drives the endothermic steam reforming reactions, and in a large unit that is well insulated, allows the process to be autothermal.

This technology was originally developed under a DOE Phase II SBIR project "Hydrogen for Refineries" (Contract DE-FG02-08ER85135), where we demonstrated syngas ($CO + H_2$) generation from extremely heavy high sulfur (2 wt.% = 20,000 ppm) hydrocarbon streams using a single stage reactor. We demonstrated over 500 hours of operation using atmospheric tower bottoms (ATB, b.p. > 650°F) and vacuum tower bottoms (VTB, b.p. > 1050°F) obtained from Valero Energy refineries (both extremely heavy and high sulfur content feedstocks), as well as biomass pyrolysis oil, NorPar 12 (a 50:50 mixture of C_{11} and C_{12} alkanes), vacuum gas oil (VGO) and other refined fuels such as JP-8 and methane. We then fabricated and operated a bench-scale DCFB reactor system to demonstrate the continuous reforming and regeneration reactions. Figure 2 shows steam reforming data for heavy vacuum gas oil (HVGO) as the feedstock to the bench-scale *HyRes* DCFB reactor system. The hydrogen content of the syngas was about 70% on a dry basis, which is typical regardless of the hydrocarbon feedstock because the syngas is essentially at thermodynamic equilibrium at these temperatures (ca. 850°C). Downstream processes such as water gas shift (for increasing the amount of H_2 in the syngas) and pressure swing absorption (PSA) purification are well developed technologies that can be added downstream of this reactor system to tailor the syngas to specific applications or purify hydrogen.

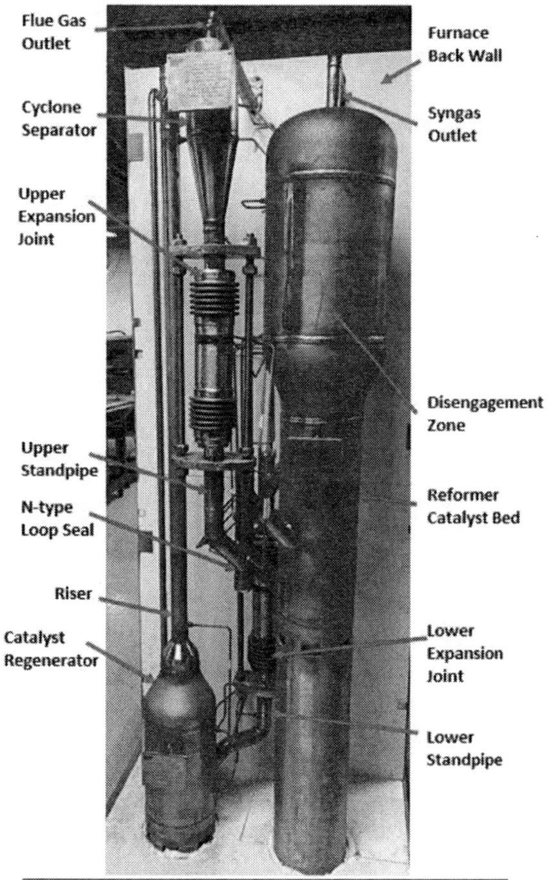

Figure 1: Pilot-scale skid-mounted dual circulating fluidized bed (DCFB) reactor system currently being assembled at TDA.

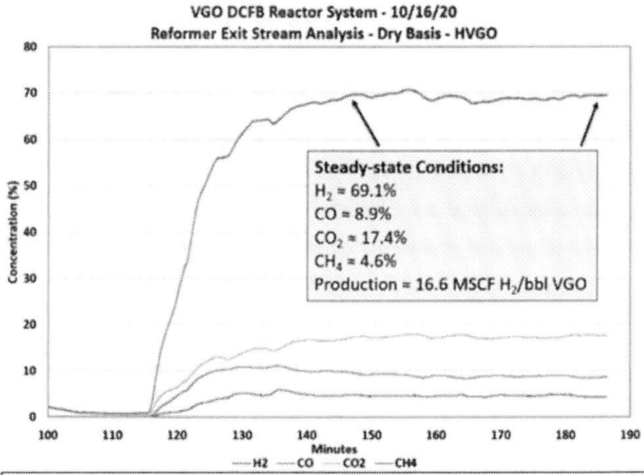

Figure 2: Steady-state analyzer data from TDA's lab-scale HyRes DCFB reactor performing steam reforming of heavy vacuum gas oil (HVGO) (DOE Phase II project).

551b Fast Start-up, Fuel Flexible Cpox Reformer for Efficient Syngas Generation from Sustainable Feedstocks

Hani Hawa, Subir Roychoudhury, Saurabh A. Vilekar, Christian Junaedi, John Hong

Precision Combustion, Inc. (PCI) has demonstrated a Catalytic Partial Oxidation (CPOX) reforming approach for low carbon intensity hydrogen (H_2) production from sustainable feedstocks, including Synthetic liquid fuels (Synfuels). Iso-Paraffinic Kerosene (IPK) and Hydro processed Esters and Fatty Acid based Synthetic Paraffinic Kerosene (HEFA-SPK) fuels were examined. HEFA-SPK is a hydrocarbon-based renewable biofuel that is chemically equivalent to fossil diesel fuel and can be used as a drop-in replacement for aviation fuels. Test data indicated LHV-based reforming efficiency of >95% of the equilibrium value and stable operation without coking. We have been developing CPOX and other application-specific fuel reforming approaches with both liquid fuels, such as jet fuel and diesel, and other sustainable fuels.

Our CPOX designs are based on a catalyzed short contact time, high surface area support (Microlith®), patented and manufactured by PCI. It reforms fuels to syngas in millisecond timeframes. No water or steam is needed for the fuel reformation yet coking is avoided. The short channel length, high cell density, metallic Microlith support allows exceptionally high heat and mass transfer as well improved mixing with minimal pressure drop. These ensure short startup time and fast transient response. This also permitted a compact CPOX reforming. It operates at reaction temperatures and stoichiometries that are favorable for avoiding high temperatures while operating at high selectivity to syngas (i.e., carbon formation avoidance) and overcomes challenges with alternative substrates and catalysts. These characteristics make it suitable and attractive for integration with advanced power conversion devices, e.g., solid oxide fuel cells (SOFCs).

We will report data from sub-scale testing with IPK and HEFA-SPK and compare results with equilibrium predictions as well as performance data from state-of-the-art syngas generation via other approaches.

551c Maximizing Conversion and Coking Stability for Hydrogen Production Via Dry Reforming of Methane over a Ni/Zeolite Catalyst

Janet Baffoe, Xuan Pham, Corey Leclerc

Global energy demand keeps rising due to increasing human population. As a result, there is a need to find alternative ways of producing energy for human use aside from fossil fuels. Hydrogen generation from biomass is one effective and sustainable way to contribute to increasing energy production. Although different processing techniques are being used recently, dry reforming remains an interesting innovative technique as it utilizes two greenhouse gases (Methane and Carbon dioxide), thus contributing to lowering carbon emissions. Deactivation, stability, cost, and selectivity are some of the primary difficulties that heterogeneous catalysis for dry reforming of methane (DRM) faces. A more competent way of producing catalysts is now ongoing in recent catalysis work.

The performance of catalysts used for the dry reforming of methane strongly depends on the synthesis method, selection of active metals and supports as well as reaction conditions. Acid and basic sites of catalysts also play an important role to reduce coking and enhance catalyst stability. If the catalyst is more basic, it improves the activation of CO2 which is acidic, thereby reducing carbon deposits on the catalyst, which are the primary mode of catalyst deactivation. My research focus is on using low cost and effective Ni metal catalysts and zeolite as a support. Specifically, I am using a clinoptilolite zeolite mined in Winston, NM. In general, zeolites have good properties such as high surface area, high porous structure for cations, and good stability at higher temperatures. They are mostly used for removing heavy metals from wastewater, and also as a catalyst for other pharmaceutical and petrochemical processes. Though some work has been done using zeolites as supports for dry reforming of methane, this is the first time the Winston Zeolite has been investigated as a catalyst support.

A 5% Ni/zeolite catalyst was synthesized using incipient wet impregnation, after which thermo-gravimetric analysis was carried out to know at which temperature the nitrate will decompose to oxides. Temperature programmed desorption was performed using hydrogen to reduce the oxides to get metallic Ni on zeolite. The temperature at which the NiO reduced to Ni was at 450oC. To know how well Ni metal is dispersed on zeolite, Hydrogen chemisorption was carried out to compare zeolite support with Al2O3 and CeO2 after TPD. According to results, zeolite support showed higher dispersion of Ni metal with smaller active particle diameter as compared with Al2O3 and CeO2. The acidic and basic sites of zeolite were analyzed using CO2 - TPD within a temperature range of 35oC to 600oC. Highest peak was seen at a temperature around 600oC indicating that zeolite has a strong basic site. NH3 - TPD was also conducted within a temperature range of 35oC to 600oC, with a low weak acidic peak at the initial temperature, indicating that it has weak acidic sites.

Future plans will be to react this catalyst with CH4 and CO2 to produce H2 and CO at 600°C - 800°C. Other analysis that will be conducted on the catalyst includes; surface area determination using BET, XRD for crystal and structure of catalyst, Pyridine FTIR analysis to determine types of acidic sites, and Temperature-Programmed Oxidation to know the amount and type of carbon formed on the catalyst. Alternate way of synthesizing catalyst will also be compared with the incipient wet impregnation method. This will help us modify and improve on the catalyst properties to ensure low deactivation, and high efficiency.

551d High Conversion of Two Major Greenhouse Gases to Syngas through Dry Reforming Reaction over Efficient Nickel-Hydroxyapatite Catalyst

Hanaa Hassini, Esma Ines Achouri

While emissions of greenhouse gases are increasing in the atmosphere due to human activities and industrial development which focuses mainly on fossil fuels as main source of energy. Efforts have been made to valorize these gases composed of carbon dioxide and methane to be used as reactants to produce syngas which is a mixture of CO and H_2 following the dry reforming reaction (DRM). However, one of the main challenges of DRM reaction is catalyst deactivation caused by coke deposition. The synthesis of a coke-resistant catalyst is a crucial step in DRM to ensure a good catalytic performance. Nickel-based catalysts are tested on Hydroxyapatite support. These catalysts were prepared by the wet impregnation method. The dry reforming reaction runs were carried out in a quartz fixed bed reactor at 800°C, atmospheric pressure, and different values of gas hourly space velocity (GHSV). CH_4 and CO_2 were emulsified with CH_4/CO_2 ratio equal to one. The results revealed high conversion reaching 93% for CH_4 and 98% for CO_2 at a low value of GHSV. Different characterization techniques were used to evaluate the physical and chemical properties of the fresh and spent catalysts such as: X-ray diffraction (XRD), electron microscope (SEM) coupled with energy-dispersive X-ray (EDX), temperature programmed reduction (TPR-H_2), thermogravimetric analysis (TGA) coupled with Gas chromatography mass spectrometry (GC-MS), Brunauer-Emmet-Teller (BET) and Barrett Joyner-Halenda (BJH) method. To sum up, Nickel based catalysts over hydroxyapatite showed a good performance and a good thermal stability. This later could be a key to minimize coke deposition which causes catalysts deactivation and eventually reduce the conversion of methane and carbone dioxide.

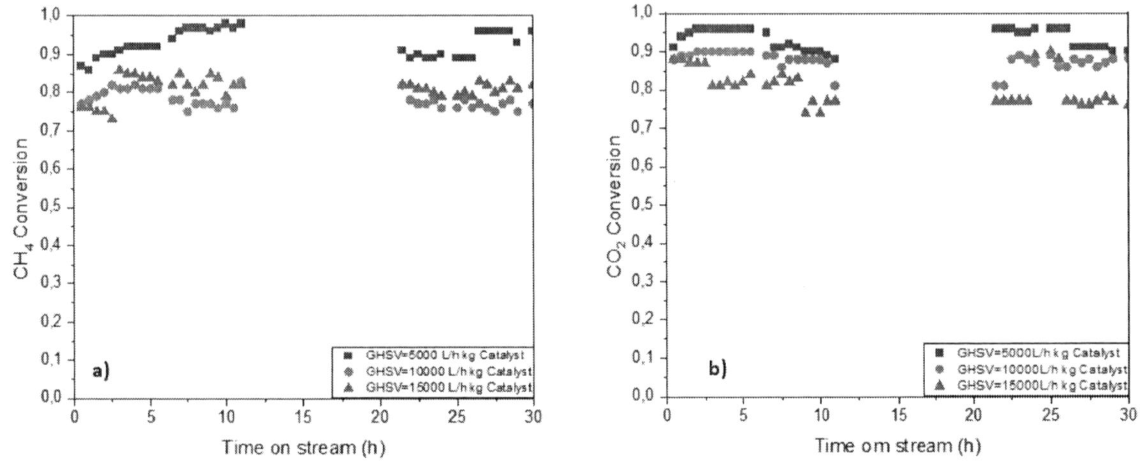

551e Dry Reforming of Methane over Ni Catalysts on High Oxygen Ion Conductivity Supports

Xuan Pham, Janet Baffoe, Corey Leclerc

The dry reforming of methane (DRM) is a chemical process that converts two greenhouse gases, methane and carbon dioxide, to syngas (a mixture of hydrogen and carbon monoxide). In an ideal case, each mole of methane and carbon dioxide produces 2 moles of carbon monoxide and hydrogen. This reaction is extremely highly endothermic and requires high operating temperatures. A H2/CO molar ratio of 1 is an ideal feed for synthesizing hydrocarbons via Fischer-Tropsch or oxygenated chemicals. Biogas is an attractive feedstock for DRM. Despite environmental and economic benefits, DRM is still not an industrial process due to the rapid catalyst deactivation caused by coke formation and sintering under severe operating conditions.

This research evaluates the effect of support materials on the performance of Ni-based catalysts, especially supports with high oxygen conductivity. Nickel is used as an active metal because of its competitive catalytic activity and affordable costs. Ceria (CeO2) is an interesting support because of its ability to form and eliminate oxygen vacancies, giving it a high oxygen ion conductivity. The oxygen vacancies on the surface of this oxide facilitate the adsorption and dissociation of CO2, improve reforming activity, and leave mobile oxygens on the catalyst surface. These mobile oxygens will react with carbon formed on the catalyst, release carbon monoxide or other carbonate species, significantly reduce coke formation, and improve the catalyst life. Previous work suggests doping CeO2 with rare earth elements, such as Gadolinium, enhances its oxygen ion conductivity. Besides Gadolinium Doped Ceria (GDC), Yttrium Stabilized Zirconia (YSZ) also possesses high oxygen ion conductivity. So, in this work, we investigate three supports with high oxygen ion conductivity and one with low oxygen ion conductivity (Alumina Al2O3) for their activity and coke resistance. Besides oxygen ion mobility, the active metal particle size and acidity/basicity of support materials have also been found to play a decisive role in the performance of DRM catalysts, especially with respect to coke formation. Carbon formation increases with particle size of nickel, so catalysts with high metal dispersion or small metal particles have strong resistance to coking. Acidic support materials tend to speed up methane cracking by acid sites and lead to the rapid deactivation of the catalysts. Basicity, in contrast, is treated as a desirable material property, allowing enhanced carbon dioxide adsorption, reduced activity of methane decomposition, and decreased carbon deposition. Therefore, during our experiments, these properties of supports and catalysts are characterized along with their catalyst activity, stability, and deactivation.

551f Syngas Production from Methane Steam Reforming over Nickel Supported Hydroxyapatite Catalyst

Sabrina Bahia Karakache, Esma Ines Achouri, Nicolas Abatzoglou

Syngas is a very important intermediate in chemical industry for energy chemicals production through F–T synthesis. Methane steam reforming (MSR) is a well-established commercial process for syngas production. The produced syngas is in a high H_2/CO ratio close to 3 for MSR. The high endothermicity of the reforming reaction requires operating temperatures that are higher than 700°C and a H_2O/CH_4 ratio higher than 1 to reduce the coke formation. Among the various catalysts which have been used for MSR, Nickel catalyst is the most widely used for its high activity and low cost. However, coking formation and sintering of nickel particles are two main challenges for the catalyst performance.

The main objective of the present work is to study the performance and the stability of nickel supported on a sustainable bio-ceramic material which is hydroxyapatite (HAp), as a catalyst for methane steam reforming. Hydroxyapatite could be an efficient support for the MSR catalyst, due to its promising properties, which are it's thermal stability, decomposing only at a temperature above 1000°C, and its basic character.

The influence of the temperature, gas hour space velocity (GHSV) and H_2O/CH_4 ratio were investigated for the performance of Ni/HAp catalyst with different %wt loadings. The catalysts were prepared by the incipient wetness impregnation method and were analyzed by the ICP, XRD, TGA-DTA and BET techniques. The reactions were performed in a fixed bed reactor at atmospheric pressure, in a GHSV between 12000 and 20000 $ml.g_{cat}^{-1}.h^{-1}$, a temperature between 700 and 800°C, using different ratio of H_2O/CH_4 (1,1.5,1.7,2).

Methane conversion and hydrogen yield conversion reached up to 88 and 80% at a ratio of $H_2O/CH_4=1$, respectively. The activity of the 10Ni/HAp was stable for a long reaction time of 100 h, at the reaction conditions of 800°C, GHSV=16000 $ml.g_{cat}^{-1}.h^{-1}$ and $H_2O/CH_4=2$. The thermal analysis revealed that HAp is a very stable material, a total weight loss of 4.5 % only was observed. The XRD patterns showed the presence of pics related to the nickel oxide phase (NiO) and the metallic phase (Ni°) for the fresh and used catalysts, respectively. The initial support presents the hexagonal closed packed phase structure of HAp. The BET results showed that the deposition of nickel particles on the support surface decreases slightly its initial specific surface area from 60 to $50 m^2.g^{-1}$, while after the reaction, the specific surface areas decrease to 4-8 $m^2.g^{-1}$ which can be a sign of the particles sintering.

551g Crude Methanol from Alternative Energy Source As Low-Cost, Low-Carbon Hydrogen Carrier

Hsiang-Sheng Chen, Kritika Sharma, Vijay Ramani, Nazim Muradov, Paul Yelvington, Edwin Yik

Methane emissions from stranded associated gas, biomass digesters, and municipal wastes can be avoided and instead converted to methanol from a gas-to-methanol plant on an easily deployable commercial trailer [1]. This methanol from otherwise flared or emitted methane can be a safer hydrogen medium than ammonia, and a more economic hydrogen medium for mid-to-long-range transportation relative to hydrogen in the form of compressed gas, cryogenic liquid, and metal hydrides [2]. Moreover, methanol synthesized from these unconventional sources can be used as a low-carbon hydrogen carrier, which can be electrolyzed to form hydrogen at the point of use. The high-purity CO_2 byproduct stream can be captured to further decrease the carbon intensity (CI).

Compared to conventional methanol from pipelined natural gas, alternative methanol, as a hydrogen carrier, is attractive due to its decentralized production, which lends access to cheap gas and energy independence for regional networks. The decreasing cost of renewable electricity further provides an incentive for electrolysis. Such a process, however, is more economically viable if crude methanol can be directly fed to electrolyzers, bypassing any requisite upgrading through costly distillation processes at more centralized facilities. From our estimation, methanol upgrading steps account for more than 10% of the overall cost of decentralized methanol; direct use of crude methanol thus avoids these costs and further decreases the CI of this alternative methanol by more than 10%. The tradeoff involves the energy penalty associated with the loss in electrolysis efficiency resulting from the presence of trace impurities within crude methanol. While inorganic impurities, such as chloride and sulfur species, can be managed or removed by upstream cleaning processes, organic impurities that persist as natural byproducts of methanol synthesis reactions are challenging to avoid [3]. The removal of ethanol, the most common organic reaction byproduct, is particularly energy intensive because of its similar volatility to methanol. These higher alcohols are known to poison catalysts for methanol electrolysis and lower the efficiency of hydrogen production.[4] Feeding crude methanol into a point-of-use electrolyzer makes economic sense when the electrochemical energy penalty associated with the presence of these organic impurities is lower than the cost of transport and distillation of crude methanol to grade methanol.

In this work, sensitivity tests are performed by bench-scale electrochemical reactions to quantify the impact of trace (0 to 5000 ppm) mixed C_2-C_4 alcohols on methanol electrolysis simulating the composition of actual crude methanol produced from alternative resources. The experiment data is then used in techno-economic assessment and the learning can be used as a decision-making tool to determine the conditions and scenarios under which direct electrolysis of crude methanol is economically feasible.

Reference

1. Yelvington, P. E.; Browne, J. B.; Yik, E.; Merical, K. I.; Dean, A. J.; Randolph, A. On Piston Engines as Hydrocarbon Gas Reformers for Modular, Distributed Chemical Production, *JAECS,* 13, **2023,** 100117
2. Papadias, D.; Peng, J-K, Ahluwalia, R. K. Hydrogen carriers: Production, transmission, decomposition, and storage, *J. Hydrog. Energy,* 46, **2021**, 24169-24189
3. Pontzen, F.; Liebner, W.; Gronemann V.; Rothaemel M.; Ahlers B. CO_2-based methanol and

DME – Efficient technologies for industrial scale production, *Catalysis Today*, 1, **2011**, 242-250

4. Tsujiguchi, T.; Furukawa, T.; Nakagawa, N. Effect of the impurities in crude bio-methanol on the performance of the direct methanol fuel cell, *J. Power Sources*,196, **2011**, 9339-9345

551h Energy-Saving Electrochemical Hydrogen Production Coupled with Value-Added Isobutanol Oxidation to Isobutyric Acid

Ruiqi Du, Siqi Zhao, Kaizheng Zhang, Shuairen Qian, Yi Cheng

The wide application of electrochemical hydrogen production is hindered by the high energy consumption of conventional water electrolysis. One of the key reasons lies in the high overpotential caused by the sluggish kinetics of anodic Oxygen Evolution Reaction (OER). In addition, the low-value oxygen product cannot be utilized effectively, thus increasing the hydrogen production costs. Replacing OER with thermodynamically more favorable organics oxidation reactions can reduce electrolysis voltage. More importantly, many anodic organics oxidation reactions can produce value-added chemicals under mild condition without any external oxidant.

Isobutanol is a kind of high energy density biofuel which can be produced by the fermentation of cellulose. Isobutyric acid, as the oxidation product of isobutanol, is an important chemical intermediate which is widely used in food, fragrance, and pharmaceutical industry. Hence, the selective transformation of isobutanol to isobutyric acid via electrochemical oxidation is a promising way to upgrade biomass to valuable fine chemicals.

In this work, we demonstrate the feasibility of coupling anodic isobutanol value-added oxidation with cathodic hydrogen evolution reaction. The Nickel-based anode electrocatalyst prepared by a facile electrodeposition method can achieve high selectivity and faradic efficiency (>90%) toward isobutyric acid. Compared with conventional water electrolysis in alkaline media, the introduction of isobutanol can reduce the cell voltage by ≈200mV under the same current density. Moreover, the formation of oxygen can be avoided, improving hydrogen purity and process safety. In this way, the electrochemical reactor design can be simplified to membrane-less structure. In conclusion, we design an energy-saving electrochemical system for coproduction of hydrogen and value-added chemicals from biomass. Combing renewable energy storage and electrified fine chemicals production, the system is applicable for distributed scenarios.

507a TEA and LCA of Renewable Natural Gas: Upgrading Biogas from Anaerobic Co-Digestion of Prairie Grass with Hydroxycinnamic Co-Products

Katherine Wild, Mark Mba Wright, Lisa Schulte

Anaerobic digestion (AD) is a biological process where organic material is broken down by microorganisms to form biogas. Biogas can be burned for heat, electricity, or upgraded further using various separating technologies to form a stream composed of methane known as renewable natural gas (RNG). RNG is injected into the natural gas pipeline for general consumption. In this study, prairie grass and manure are co-digested together on a farm-site.

This study hopes to determine if hydroxycinnamic acids (HCAs) are a potential co-product for RNG. HCAs are acids derived from chemically pretreating biomass. In a previous study HCAs derived from corn stover were determined to lower the price of ethanol by $0.26/L. A previous laboratory study created a process of pretreating prairie grass in a KOH solution. The biomass co-digested with manure was found to increase the amount of methane within the biogas stream.

This study utilizes a techno-economic analysis (TEA) to determine if HCAs are a viable co-product for RNG. This process model was created in BioSTEAM to determine a minimum fuel selling price (MFSP) of RNG with HCAs as a co-product. A life cycle assessment (LCA) was also completed in openLCA to determine the effects this process has on the environment. Sensitivity and uncertainty analyses were also considered to find the MFSP and greenhouse gas (GHG) emissions of RNG under different process adaptations.

507b Low-Dimensional Techno-Economic Cost Model for Expedited Hydrothermal Liquefaction Process Evaluation

Muntasir Shahabuddin, Eduardo Italiani, Andrew R Teixeira, Nikolaos Kazantzis, Michael T. Timko

Hydrothermal liquefaction (HTL) is rapidly emerging as a method to convert abundant organic wastes into renewable fuels. Deployment of this promising technology would benefit greatly from generalized methods of predicting costs, since current methods for estimating HTL costs requires a time and resource intensive process design and techno-economic analysis (TEA) for each specific case. To enable more efficient allocation of resources, there is a need for a simple and universal tool that can predict the economic viability of HTL processes. In this study, we have developed a model that utilizes three key variables - feedstock cost, biocrude yield, and process scale - to predict the minimum selling price of fuel (MFSP) produced by an HTL process. Our model is based on empirical observations from a wide range of feedstock costs, biocrude yields, and process scales obtained through detailed TEA. By fitting adjustable parameters to this model, we were able to predict independently published values of MFSP ($N = 25$) with an accuracy of ±20.4%, as shown in Figure 1. Model predictions are within 40% of the published value 95% of the time and in the worst case the discrepancy is 45.9%, meaning that the accuracy of the universal model is sufficient for screening purposes.

Figure 1: Parity performance of model-predicted minimum fuel selling price (MFSP) against literature values of MFSP.

The resulting model is capable of predicting MFSP using readily available data inputs, including feedstock cost, process scale, and experimental results for biocrude yield, enabling order-of-magnitude estimates for HTL ventures while foregoing the resource-intensive process of performing a traditional economic analysis.

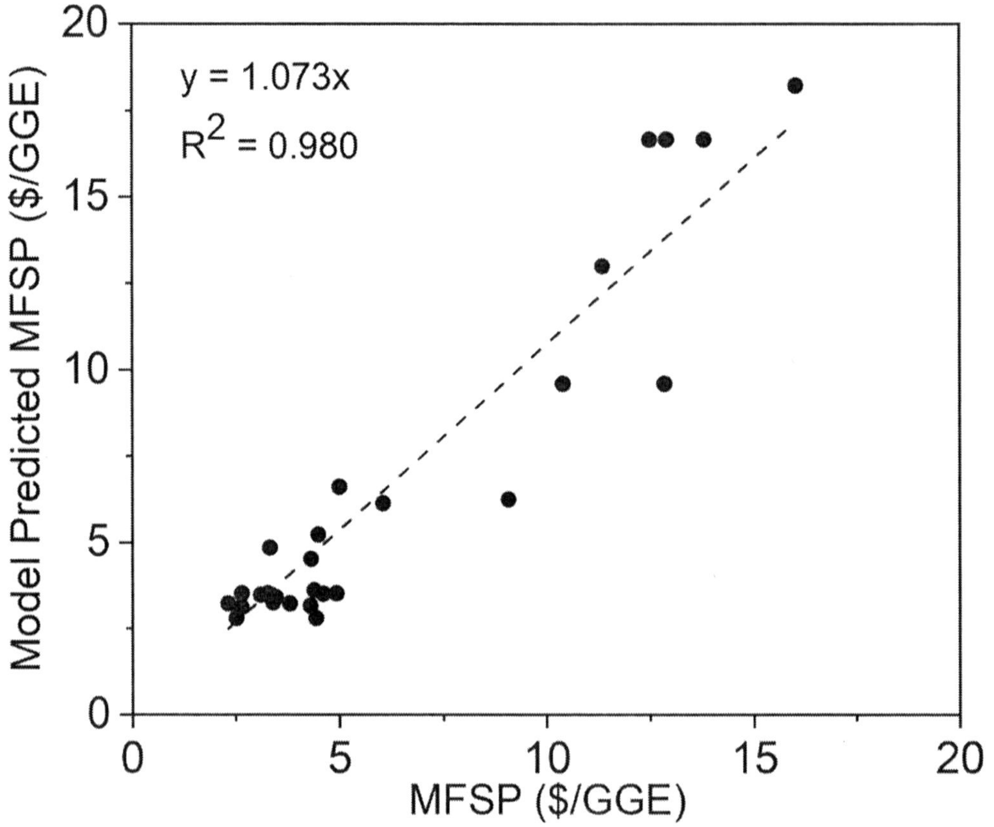
y = 1.073x
R2 = 0.980
Model Predicted MFSP ($/GGE)
MFSP ($/GGE)
0
5
10
15
20
0
5
10
15
20

507c Catalytic Hydrodeoxygenation of Hydrothermal Liquefaction (HTL) Derived Bio-Oil from Corn Stover for Sustainable Aviation Fuels

Zoe Fickbohm, Bharathkiran Maddipudi, Rajesh Shende, Khang Huynh

Hydrothermal liquefaction (HTL) is an effective thermochemical technology to convert biomass into fuels and value-added products. In this study, preprocessed corn stover was hydrothermally liquified at subcritical temperatures to generate bioproducts such as char, heavy bio-oil (HBO), liquid biocrude and gases. The extracted HBO using polar aprotic (acetone) solvent was dried and subjected to catalytic hydrodeoxygenation (HDO) for upgradation in the custom-made 20 ml SS316 reactor. HBO was characterized using gas-chromatography mass spectrometry (GC-MS) to analyze the hydrocarbons. Elemental analysis of HBO and upgraded oil was performed to understand the oxygen content and higher heating value (HHV). ASPEN plus was utilized to determine the theoretical amount of hydrogen required to perform the HDO reaction. Yields of HTL derived HBO, upgraded oil, and ASPEN plus modelling results will be presented under different experimental conditions of catalyst loading, temperature/pressure conditions, and residence time.

507d TEA/LCA of Integrated Processing of Corn Stover for Fuels and High Value Products

Bharathkiran Maddipudi, Khang Huynh, Vinod Amar, Katelyn Shell, Anuj Thakkar, Anuradha Shende, John E. Aston, Sandeep Kumar, Ram Gupta, Rajesh Shende

Worldwide, at least ½ billion-ton corn stalk or stover is generated per year. An integrated processing approach comprised of biochemical and hydrothermal technologies is developed to process corn stover into high value products that can be sold to generate revenue and offset the fuel cost. Corn bales were pre-processed at the Biomass Feedstock National User Facility (BFNUF) and the obtained pulverized corn stover (low and high ash content) were biochemically treated to obtain unhydrolyzed solids (UHS). Hydrothermal liquefaction (HTL) of UHS was performed to produce hydrochar, phenols and lactic acid. Hydrochar was further valorized into porous carbon, which was used for the energy storage application. Process engineering calculations and scale-up was performed for 1 tpd feedstock throughput. Technoeconomic analysis (TEA) and life-cycle analysis (LCA) were carried out considering optimized integrated processing. Production rate of high value-added products such as high value carbon, phenols and carboxylic acids was included in the process economics calculations. Using ASPEN plus modeling, the integrated process was scaled up to achieve a fuel cost goal of $2 gge.

2023 AIChE ANNUAL MEETING
November 5 — November 10, 2023
Hyatt Regency Orlando
Orlando, FL

Integration of Biomass Conversion and Solar Hydrogen Generation Technologies to Produce Renewable Fuels

(Principal Author)
Rahul Bhosale
Department of Chemical Engineering,
University of Tennessee, Chattanooga, TN, 37421, USA

(Co-Authors)
Sayma Akhter
Department of Chemical Engineering,
Qatar University, Doha, Qatar

Gopalakrishnan Kumar
Department of Chemistry, Life Sciences, and Environmental Technology,
University of Stavanger, Norway.

Prepared for Presentation at
American Institute of Chemical Engineers
2023 AIChE Annual Meeting
Orlando, FL, USA
November 5 – 10, 2023

Integration of Biomass Conversion and Solar Hydrogen Generation Technologies to Produce Renewable Fuels

Rahul R. Bhosale*,[1], Sayma Akther[2], Gopalakrishnan Kumar[3]

[1]Department of Chemical Engineering, University of Tennessee, Chattanooga, TN, 37421, USA
[2]Department of Chemical Engineering, Qatar University, Doha, Qatar
[3]Department of Chemistry, Life Sciences, and Environmental Technology, University of Stavanger, Norway

Keywords: Biomass Conversion, Solar Hydrogen, Renewable Fuels

Abstract

This investigation is divided into two parts: a) solar fuel production and b) biomass conversion. The solar fuel section is focused on the synthesis, characterization, and TGA evaluation of the La-Sr-Mn-Co-perovskites. The biomass section deals with cultivating the microalgae consortia in a cost-effective photo reactor.

1. Introduction

Production and utilization of alternative fuels from renewable resources such as biomass and solar energy are potential replacement options for fossil fuels[1–17]. To be specific, one of the options to reduce the production and release of CO_2 into the atmosphere is to re-utilize the CO_2 to produce fuels[18–22]. In addition to fulfilling the world energy demand, alternative fuels can also play a vital role in reducing the pressure from the research community regarding their efforts towards resolving environmental issues induced by the continuous use of fossil fuels and industrialization [23–36].

In one of our current research projects, we are focusing on integrating two processes, i.e., a) the conversion of biomass and b) hydrogen production from solar thermochemical cycles for the production of renewable fuels. In the case of part a, at different dilution concentrations as N (nitrogen) and P (phosphorous) sources, microalgae consortia are cultivated using digestate as a substrate to grow and accumulate biomass and other components. The characteristics changes during the growth from its original niche have been reported in this paper. Part b focuses on synthesizing, characterizing, and testing the La-based perovskites for solar hydrogen production via the thermochemical water splitting. Although plenty of metal oxides have been investigated for solar hydrogen and solar CO production [37–77], the solar thermochemical community is still searching for the best metal oxide combination that can sustainably produce solar fuels.

2. Solar Fuel Production

For the 1st section of the study, solar fuel production, we have synthesized La-Sr-Mn-Co-doped (LSMC) perovskites using the combustion method. For synthesis, four metal nitrates (La nitrate, Sr Nitrate, Mn Nitrate, and Co nitrates) and an oxidizer glycine were dissolved into deionized water (step I) with a fuel-to-nitrate ratio of one. The formed uniform solutions were preheated until the boiling temperature of water by maintaining a hot plate temperature above 300°C. The thick viscous glycine and metal nitrate gel were formed after completely losing water vapor (step II). Then, the gel temperature rises until auto-ignition temperature, wherein the combustion reaction propagates with the formation of desired perovskites (step III). Formed porous and soft powder of LSMC perovskites was crushed into fine nanoparticles and finally calcined at 1000°C (heating rate 5°C/min) for 3h inside the muffle furnace in the presence of atmospheric air (step IV). The synthesized LSMC perovskites were analyzed using PXRD, and the results obtained indicate phase pure composition. Likewise, the SEM analysis confirmed that the combustion synthesis yielded a porous morphology. The synthesized LSMC perovskites' redox reactivity was examined using a high-temperature thermogravimetric analyzer purchased from SETARAM, France (Fig. 1).

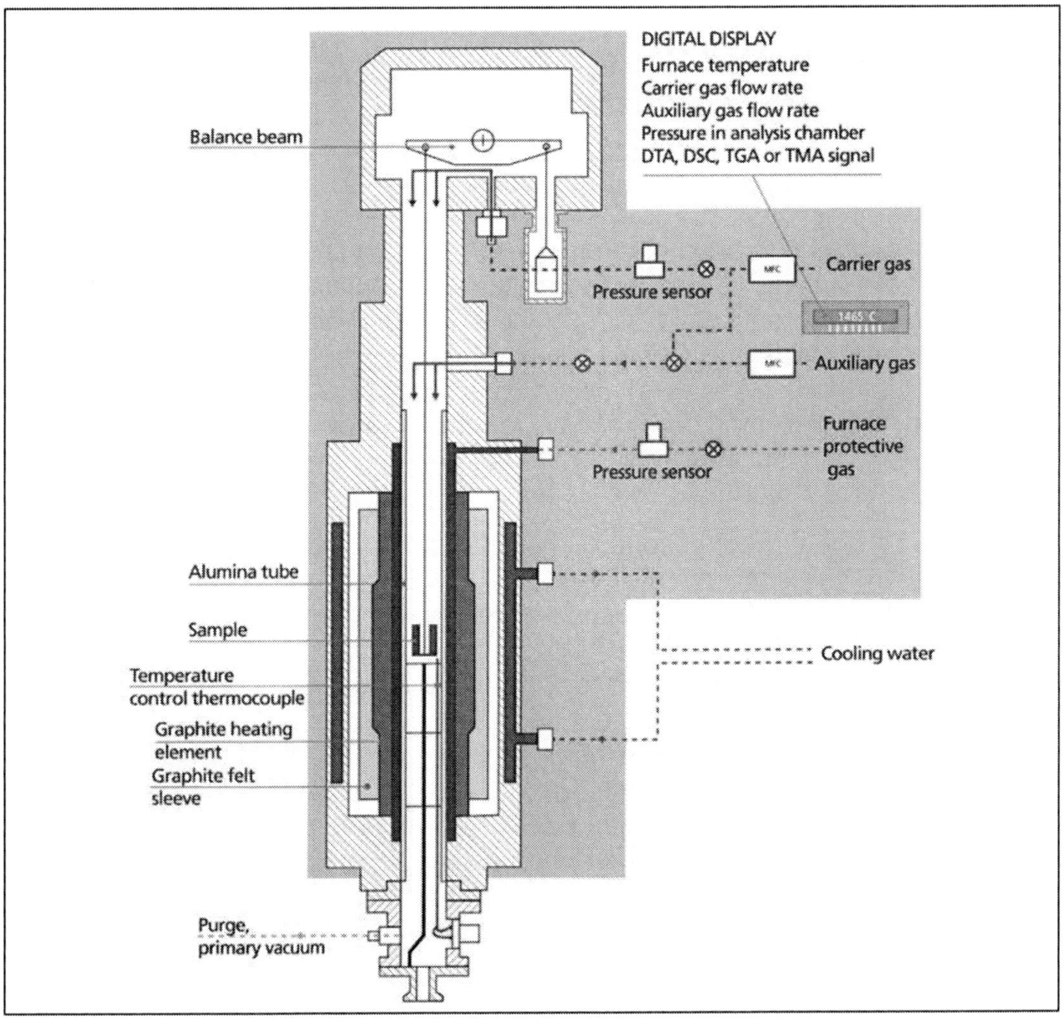

Fig. 1. Setaram Setsys TGA instrument utilized to conduct the thermochemical experiments.

The obtained TGA plot of each LSMC perovskite was regularised to remove the mass measurement error due to buoyancy. This TGA experiment was conducted with an empty crucible under an identical input temperature program, the number of cycles, reaction time, and gas flow rate. The normalized percentage of mass variation (after eliminating the buoyancy effect) of each LSMC perovskite was used for the performance comparison. The mass changes plot of the single cycle (TR and CS) of LSMC perovskites with a thermal reduction temperature of 1400°C and oxidation temperature of 1000°C is shown in Fig. 2. It can be seen that each LSMC perovskite mass decreases with the rise in temperature and during TR step (at 1400 °C for 60 min) due to release of lattice O_2 from each LSMC nanoparticles. During the second CO_2 conversion step, reduced LSMC perovskite oxidized with the formation of CO at a low temperature of 1000°C for a reaction period of 30 min with a feed gas mixture composition of 50% CO_2 and 50% Ar. In the oxidation reaction, CO_2 splits and releases CO and $0.5O_2$. The lattice structure of reduced LSMC perovskites absorbs this released $0.5O_2$ at a lower temperature of 1000°C.

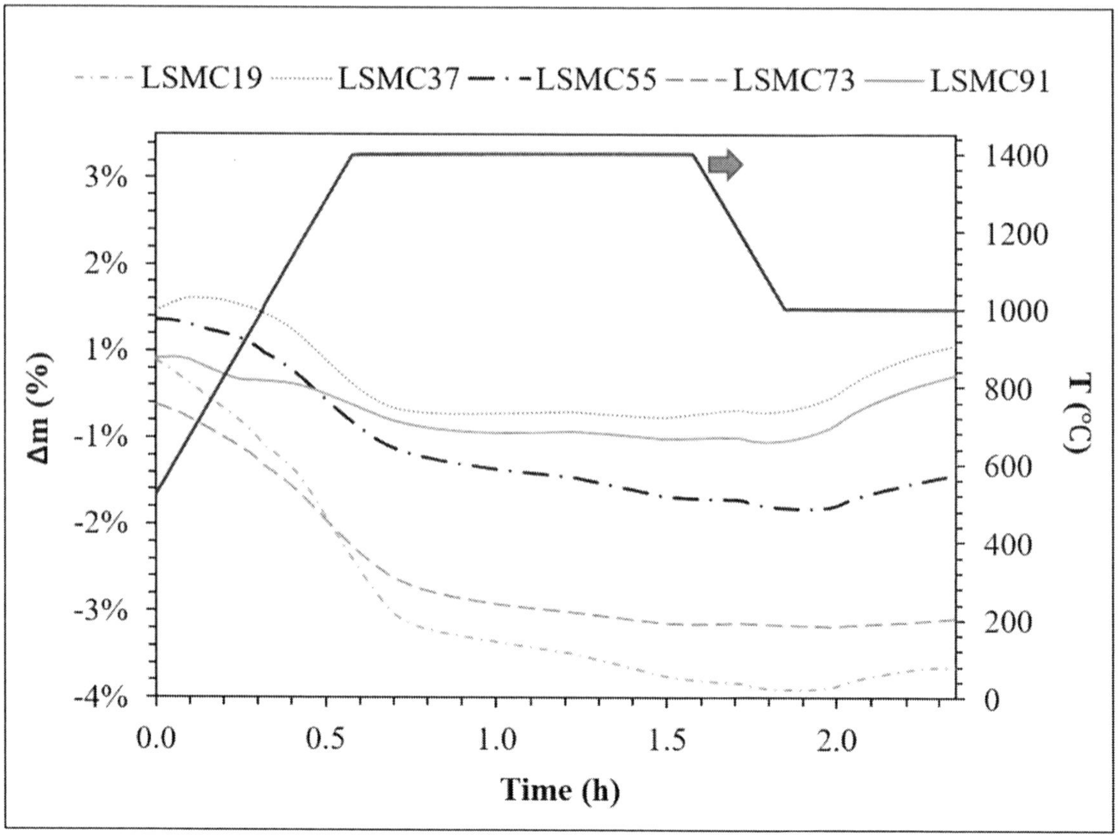

Fig. 2. Setaram Setsys TGA plots for LSMC perovskites.

3. Biomass Conversion

Microalgae consortia are cultivated using digestate as substrate at different dilution concentrations as N (nitrogen) and P (phosphorous) sources to grow and accumulate biomass and other components. Three digestate concentrations were prepared using water as a diluting agent and are denoted as I (low), II (medium), and III (high, almost original concentration). After a cultivation period of >30 days, the N and P concentrations achieved 100% removal (reached more than 95% within 15 days of cultivation). P was entirely removed after 15 days due to its low concentration in the seed source. The biomass productivity of VS (volatile solids), carbohydrate, protein, and lipid production has been determined weekly. The pH variation was not relative to the nutrient depletion and accumulation of lipids. The OD (optical density) values reached nearly 7.5 and 6.4 for the absorbance at 680 and 750 nm, respectively. Additionally, carbohydrate, protein, and lipid content were enriched in low, high, and medium concentrations, respectively. Thus, this study proves the cost-effective strategy for cultivating microalgae consortia, which could be further converted to biofuels, and the waste/wastewater toxic reduction as dual benefits. The microalgae consortia, prospected from various ecological niches, have been cultivated in cost-effective photo reactors (10 L capacity and 5 L as working volume, plastic bags) for about 40 days. The characteristics changes during the growth from its original niche have been provided in Table 1.

Table 1. Characteristics of digestate and mixed microalgae consortia.

Parameters	**Unit**	**digestate**	**Microalgae consortia**
pH	-	6.8±0.4	7.3±0.6
tCOD	g/L	10.6±1.0	1.3±0.2
sCOD	g/L	1.7±0.3	0.2±0.04
Total Nitrogen	mg/L	314±15	ND
Total Phosphorous	mg/L	16±1.2	ND
NH4-N	mg/L	284±18	ND
PO3-P	mg/L	25±3	ND
Total Solids	g/L	13.8±0.8	1.3±0.2
Volatile Solids	g/L	4.6±0.6	0.6±0.1

4. Conclusions

In the case of solar fuel production, LSMC perovskites were successfully synthesized via combustion and characterized using PXRD and SEM. The TGA analysis of the derived perovskites was also conducted in the temperature range of 1000 to 1400°C. In biomass conversion, the microalgae consortia, prospected from various ecological niches, have been cultivated in cost-effective photo reactors (10 L capacity and 5 L as working volume, plastic bags) for about 40 days.

5. Acknowledgments

The authors would like to acknowledge the support provided by Qatar University International Research Collaboration Co-funds Grant No. IRCC-2021-002 and the Ruth S. Holmberg Grant for Faculty Excellence, University of Tennessee, Chattanooga, TN, USA.

References

[1] Almomani F, Bohsale RR. Optimizing nutrient removal of moving bed biofilm reactor process using response surface methodology. Bioresour Technol 2020;305:123059. https://doi.org/10.1016/j.biortech.2020.123059.

[2] Tungal R, Shende R V. Hydrothermal liquefaction of pinewood (Pinus ponderosa) for H2, biocrude and bio-oil generation. Appl Energy 2014;134:401–12. https://doi.org/10.1016/j.apenergy.2014.07.060.

[3] Pugazhendhi A, Nagappan S, Bhosale RR, Tsai P-C, Natarajan S, Devendran S, et al. Various potential techniques to reduce the water footprint of microalgal biomass production for biofuel—A review. Science of The Total Environment 2020;749:142218. https://doi.org/10.1016/j.scitotenv.2020.142218.

[4] Rodionova MV, Poudyal RS, Tiwari I, Voloshin RA, Zharmukhamedov SK, Nam HG, et al. Biofuel production: Challenges and opportunities. Int J Hydrogen Energy 2017;42:8450–61. https://doi.org/10.1016/j.ijhydene.2016.11.125.

[5] Shurair M, Almomani F, Bhosale R, Khraisheh M, Qiblawey H. Harvesting of intact microalgae in single and sequential conditioning steps by chemical and biological based – flocculants: Effect on harvesting efficiency, water recovery and algal cell morphology. Bioresour Technol 2019;281:250–9. https://doi.org/10.1016/j.biortech.2019.02.103.

[6] Voloshin RA, Rodionova M V., Zharmukhamedov SK, Nejat Veziroglu T, Allakhverdiev SI. Review: Biofuel production from plant and algal biomass. Int J Hydrogen Energy 2016;41:17257–73. https://doi.org/10.1016/j.ijhydene.2016.07.084.

[7] Rajesh Banu J, Yukesh Kannah R, Kavitha S, Ashikvivek A, Bhosale RR, Kumar G. Cost effective biomethanation via surfactant coupled ultrasonic liquefaction of mixed microalgal biomass harvested from open raceway pond. Bioresour Technol 2020;304:123021. https://doi.org/10.1016/j.biortech.2020.123021.

[8] Matin MA, Kumar A, Bhosale RR, Saleh Saad MAH, Almomani FA, Al-Marri MJ. PdZn nanoparticle electrocatalysts synthesized by solution combustion for methanol oxidation reaction in an alkaline medium. RSC Adv 2017;7:42709–17. https://doi.org/10.1039/C7RA07013F.

[9] Almomani F, Shawaqfah M, Bhosale RR, Kumar A, Khraisheh MAM. Intermediate ozonation to enhance biogas production in batch and continuous systems using animal dung and agricultural waste. Int Biodeterior Biodegradation 2017;119:176–87. https://doi.org/10.1016/j.ibiod.2016.11.008.

[10] Abanades S, Abbaspour H, Ahmadi A, Das B, Ehyaei MA, Esmaeilion F, et al. A critical review of biogas production and usage with legislations framework across the globe. International Journal of Environmental Science and Technology 2022;19:3377–400. https://doi.org/10.1007/s13762-021-03301-6.

[11] Almomani F, Bhosale RR, Khraisheh MAM, Shawaqfah M. Enhancement of biogas production from agricultural wastes via pre-treatment with advanced oxidation processes. Fuel 2019;253:964–74. https://doi.org/10.1016/j.fuel.2019.05.057.

[12] Kabeyi MJB, Olanrewaju OA. Biogas Production and Applications in the Sustainable Energy Transition. Journal of Energy 2022;2022:1–43. https://doi.org/10.1155/2022/8750221.

[13] Rodriguez JA, Liu P, Stacchiola DJ, Senanayake SD, White MG, Chen JG. Hydrogenation of CO $_2$ to Methanol: Importance of Metal–Oxide and Metal–Carbide Interfaces in the Activation of CO $_2$. ACS Catal 2015;5:6696–706. https://doi.org/10.1021/acscatal.5b01755.

[14] Ashok A, Kumar A, Bhosale R, Saleh Saad MA, AlMomani F, Tarlochan F. Study of ethanol dehydrogenation reaction mechanism for hydrogen production on combustion synthesized cobalt catalyst. Int J Hydrogen Energy 2017;42:23464–73. https://doi.org/10.1016/j.ijhydene.2017.01.175.

[15] GO K, SON S, KIM S. Reaction kinetics of reduction and oxidation of metal oxides for hydrogen production. Int J Hydrogen Energy 2008;33:5986–95. https://doi.org/10.1016/j.ijhydene.2008.05.039.

[16] Kumar A, Ashok A, Bhosale RR, Saleh MAH, Almomani FA, Al-Marri M, et al. In situ DRIFTS Studies on Cu, Ni and CuNi catalysts for Ethanol Decomposition Reaction. Catal Letters 2016;146:778–87. https://doi.org/10.1007/s10562-016-1706-9.

[17] Li J, Kazakov A, Dryer FL. Experimental and Numerical Studies of Ethanol Decomposition Reactions. J Phys Chem A 2004;108:7671–80. https://doi.org/10.1021/jp0480302.

[18] Almomani F, Bhosale R, Khraisheh M, Kumar A, Tawalbeh M. Photocatalytic conversion of CO2 and H2O to useful fuels by nanostructured composite catalysis. Appl Surf Sci 2019;483:363–72. https://doi.org/10.1016/j.apsusc.2019.03.304.

[19] Dhabarde N, Selvaraj J, Yuda A, Kumar A, Subramanian VR. Review of photocatalytic and photo-electrocatalytic reduction of CO2 on carbon supported films. Int J Hydrogen Energy 2022;47:30908–36. https://doi.org/10.1016/j.ijhydene.2022.02.124.

[20] Ashok A, Kumar A, Bhosale RR, Almomani F, Saleh Saad MAH, Suslov S, et al. Influence of fuel ratio on the performance of combustion synthesized bifunctional cobalt oxide catalysts for fuel cell application. Int J Hydrogen Energy 2019;44:436–45. https://doi.org/10.1016/j.ijhydene.2018.02.111.

[21] Hao Y, Jin J, Jin H. Thermodynamic analysis of isothermal CO2 splitting and CO2-H2O co-splitting for solar fuel production. Appl Therm Eng 2020;166:113600. https://doi.org/10.1016/j.applthermaleng.2019.04.010.

[22] Ashok A, Kumar A, Bhosale RR, Saleh MAH, Ghosh UK, Al-Marri M, et al. Cobalt oxide nanopowder synthesis using cellulose assisted combustion technique. Ceram Int 2016;42:12771–7. https://doi.org/10.1016/j.ceramint.2016.05.035.
[23] Bhosale RR, Kumar A, AlMomani F, Ghosh U, AlNouss A, Scheffe J, et al. CO_2 Capture Using Aqueous Potassium Carbonate Promoted by Ethylaminoethanol: A Kinetic Study. Ind Eng Chem Res 2016;55:5238–46. https://doi.org/10.1021/acs.iecr.5b04398.
[24] Vaidya PD, Kenig EY. CO2-Alkanolamine Reaction Kinetics: A Review of Recent Studies. Chem Eng Technol 2007;30:1467–74. https://doi.org/10.1002/ceat.200700268.
[25] Gulied M, Al Momani F, Khraisheh M, Bhosale R, AlNouss A. Influence of draw solution type and properties on the performance of forward osmosis process: Energy consumption and sustainable water reuse. Chemosphere 2019;233:234–44. https://doi.org/10.1016/j.chemosphere.2019.05.241.
[26] Zhang QH, Yang WN, Ngo HH, Guo WS, Jin PK, Dzakpasu M, et al. Current status of urban wastewater treatment plants in China. Environ Int 2016;92–93:11–22. https://doi.org/10.1016/j.envint.2016.03.024.
[27] Mohamad S, Fares A, Judd S, Bhosale R, Kumar A, Gosh U, et al. Advanced wastewater treatment using microalgae: effect of temperature on removal of nutrients and organic carbon. IOP Conf Ser Earth Environ Sci 2017;67:012032. https://doi.org/10.1088/1755-1315/67/1/012032.
[28] Ahmed SN, Haider W. Heterogeneous photocatalysis and its potential applications in water and wastewater treatment: a review. Nanotechnology 2018;29:342001. https://doi.org/10.1088/1361-6528/aac6ea.
[29] Bhosale RR, Mahajani V V. Kinetics of thermal degradation of renewably prepared amines useful for flue gas treatment. Journal of Renewable and Sustainable Energy 2013;5. https://doi.org/10.1063/1.4831960.
[30] Moreira D, Pires JCM. Atmospheric CO2 capture by algae: Negative carbon dioxide emission path. Bioresour Technol 2016;215:371–9. https://doi.org/10.1016/j.biortech.2016.03.060.
[31] Bhosale RR, Mahajani V V. Kinetics of Absorption of Carbon Dioxide in Aqueous Solution of Ethylaminoethanol Modified with N-methyl-2-pyrolidone. Sep Sci Technol 2013;48:2324–37. https://doi.org/10.1080/01496395.2013.805228.
[32] Stolaroff JK, Ye C, Nguyen DT, Oakdale J, Knipe JM, Baker SE. CO2 Absorption Kinetics of Micro-encapsulated Ionic Liquids. Energy Procedia 2017;114:860–5. https://doi.org/10.1016/j.egypro.2017.03.1228.
[33] Almomani F, Bhosale R, Kumar A, Khraisheh M. Potential use of solar photocatalytic oxidation in removing emerging pharmaceuticals from wastewater: A pilot plant study. Solar Energy 2018;172:128–40. https://doi.org/10.1016/j.solener.2018.07.041.
[34] Toth AJ. Modelling and Optimisation of Multi-Stage Flash Distillation and Reverse Osmosis for Desalination of Saline Process Wastewater Sources. Membranes (Basel) 2020;10:265. https://doi.org/10.3390/membranes10100265.
[35] Almomani F, Bhosale RR. Bio-sorption of toxic metals from industrial wastewater by algae strains Spirulina platensis and Chlorella vulgaris: Application of isotherm,

kinetic models and process optimization. Science of The Total Environment 2021;755:142654. https://doi.org/10.1016/j.scitotenv.2020.142654.
[36] Crini G, Lichtfouse E. Advantages and disadvantages of techniques used for wastewater treatment. Environ Chem Lett 2019;17:145–55. https://doi.org/10.1007/s10311-018-0785-9.
[37] Bhosale RR, Kumar A, AlMomani F, Ghosh U, Dardor D, Bouabidi Z, et al. Solar co-production of samarium and syngas via methanothermal reduction of samarium sesquioxide. Energy Convers Manag 2016;112:413–22. https://doi.org/10.1016/j.enconman.2016.01.032.
[38] Steinfeld A. Solar hydrogen production via a two-step water-splitting thermochemical cycle based on Zn/ZnO redox reactions. Int J Hydrogen Energy 2002;27:611–9. https://doi.org/10.1016/S0360-3199(01)00177-X.
[39] Bhosale RR, Alxneit I, van den Broeke LLP, Kumar A, Jilani M, Gharbia SS, et al. Sol-Gel Synthesis of Nanocrystalline Ni-Ferrite and Co-Ferrite Redox Materials for Thermochemical Production of Solar Fuels. MRS Proceedings 2014;1675:203–8. https://doi.org/10.1557/opl.2014.866.
[40] Diver RB, Miller JE, Allendorf MD, Siegel NP, Hogan RE. Solar Thermochemical Water-Splitting Ferrite-Cycle Heat Engines. J Sol Energy Eng 2008;130. https://doi.org/10.1115/1.2969781.
[41] Bhosale RR, Kumar A, Ashok A, Sutar P, Takalkar G, Khraisheh M, et al. La-Based Perovskites as Oxygen-Exchange Redox Materials for Solar Syngas Production. MRS Adv 2017;2:3365–70. https://doi.org/10.1557/adv.2017.300.
[42] Bare ZJL, Morelock RJ, Musgrave CB. A Computational Framework to Accelerate the Discovery of Perovskites for Solar Thermochemical Hydrogen Production: Identification of Gd Perovskite Oxide Redox Mediators. Adv Funct Mater 2022;32:2200201. https://doi.org/10.1002/adfm.202200201.
[43] Takalkar G, Bhosale RR. Solar thermocatalytic conversion of CO2 using PrxSr(1−x)MnO3−δ perovskites. Fuel 2019;254:115624. https://doi.org/10.1016/j.fuel.2019.115624.
[44] Li S, Wheeler VM, Kumar A, Venkataraman MB, Muhich CL, Hao Y, et al. Thermodynamic Guiding Principles for Designing Nonstoichiometric Redox Materials for Solar Thermochemical Fuel Production: Ceria, Perovskites, and Beyond. Energy Technology 2022;10:2000925. https://doi.org/10.1002/ente.202000925.
[45] Bhosale RR. Thermodynamic analysis of Ni-ferrite based solar thermochemical H2O splitting cycle for H2 production. Int J Hydrogen Energy 2019;44:61–71. https://doi.org/10.1016/j.ijhydene.2018.03.145.
[46] Scheffe JR, Li J, Weimer AW. A spinel ferrite/hercynite water-splitting redox cycle. Int J Hydrogen Energy 2010;35:3333–40. https://doi.org/10.1016/j.ijhydene.2010.01.140.
[47] Takalkar G, Bhosale R, AlMomani F. Combustion synthesized A0.5Sr0.5MnO3-δ perovskites (where, A = La, Nd, Sm, Gd, Tb, Pr, Dy, and Y) as redox materials for thermochemical splitting of CO2. Appl Surf Sci 2019;489:80–91. https://doi.org/10.1016/j.apsusc.2019.05.284.
[48] Jacot R, Moré R, Michalsky R, Steinfeld A, Patzke GR. Trends in the phase stability and thermochemical oxygen exchange of ceria doped with potentially

tetravalent metals. J Mater Chem A 2017;5:19901–13. https://doi.org/10.1039/C7TA04063F.
[49] Luciani G, Landi G, Di Benedetto A. Syngas Production Through H2O/CO2 Thermochemical Splitting Over Doped Ceria-Zirconia Materials. Front Energy Res 2020;8. https://doi.org/10.3389/fenrg.2020.00204.
[50] Bhosale RR, Shende RV, Puszynski JA. Sol-Gel Synthesis of Ferrite Foam Materials for H2 Generation from Water-Splitting Reaction. Proceedings of NSTI–Nanotech Conference & Expo., 2010, p. 368–71.
[51] Tamaura Y, Hasegawa N, Kojima M, Ueda Y, Amano H, Tsuji M. Water splitting with the Mn(II)-ferrite–CaO–H2O system at 1273K. Energy 1998;23:879–86. https://doi.org/10.1016/S0360-5442(98)00030-9.
[52] Dardor D, Bhosale RR, Gharbia S, Kumar A, AlMomani FA. Solar carbon production via thermochemical ZnO/Zn carbon dioxide splitting cycle. Journal of Emerging Trends in Engineering and Applied Sciences 2016;7:129–35.
[53] Chambon M, Abanades S, Flamant G. Solar thermal reduction of ZnO and SnO2: Characterization of the recombination reaction with O2. Chem Eng Sci 2010;65:3671–80. https://doi.org/10.1016/j.ces.2010.03.005.
[54] Bhosale RR, Shende R V, Puszynski JA. H 2 Generation From Thermochemical Water-Splitting Using Sol-Gel Synthesized Zn/Sn/Mn-doped Ni-Ferrite. International Review of Chemical Engineering (IRECHE) 2010;2.
[55] Kaneko H, Gokon N, Hasegawa N, Tamaura Y. Solar thermochemical process for hydrogen production using ferrites. Energy 2005;30:2171–8. https://doi.org/10.1016/j.energy.2004.08.020.
[56] Shende RV, Puszynski JA, Opoku MK, Bhosale RR. Synthesis of novel ferrite foam material for water-splitting application. Proceedings of NSTI Nanotech conference & expo., 2009, p. 201–4.
[57] Scheffe JR, McDaniel AH, Allendorf MD, Weimer AW. Kinetics and mechanism of solar-thermochemical H2 production by oxidation of a cobalt ferrite–zirconia composite. Energy Environ Sci 2013;6:963. https://doi.org/10.1039/c3ee23568h.
[58] Bhosale RR, Shende R V, Puszynski JA. H 2 Generation from Thermochemical Water-Splitting Using Sol-Gel Derived Ni-Ferrite. Journal of Energy and Power Engineering 2010;4.
[59] Mostrou S, Büchel R, Pratsinis SE, van Bokhoven JA. Improving the ceria-mediated water and carbon dioxide splitting through the addition of chromium. Appl Catal A Gen 2017;537:40–9. https://doi.org/10.1016/j.apcata.2017.03.001.
[60] Bhosale R, Khadka R, Puszynski J, Shende R. H2 generation from two-step thermochemical water-splitting reaction using sol-gel derived SnxFeyOz. Journal of Renewable and Sustainable Energy 2011;3. https://doi.org/10.1063/1.3659684.
[61] Agrafiotis C, Zygogianni A, Pagkoura C, Kostoglou M, Konstandopoulos AG. Hydrogen production via solar-aided water splitting thermochemical cycles with nickel ferrite: Experiments and modeling. AIChE Journal 2013;59:1213–25. https://doi.org/10.1002/aic.13882.
[62] Bhosale RR, Kumar A, AlMomani F, Ghosh U, Sutar P, Takalkar G, et al. Effectiveness of Ni incorporation in iron oxide crystal structure towards thermochemical CO2 splitting reaction. Ceram Int 2017;43:5150–5. https://doi.org/10.1016/j.ceramint.2017.01.034.

[63] Haeussler A, Julbe A, Abanades S. Investigation of reactive perovskite materials for solar fuel production via two-step redox cycles: Thermochemical activity, thermodynamic properties and reduction kinetics. Mater Chem Phys 2022;276:125358. https://doi.org/10.1016/j.matchemphys.2021.125358.
[64] Bhosale RR, Sutar P, Kumar A, AlMomani F, Ali MH, Ghosh U, et al. Solar hydrogen production via erbium oxide based thermochemical water splitting cycle. Journal of Renewable and Sustainable Energy 2016;8:034702. https://doi.org/10.1063/1.4953166.
[65] Bhosale RR. Thermodynamic efficiency analysis of zinc oxide based solar driven thermochemical H2O splitting cycle: Effect of partial pressure of O2, thermal reduction and H2O splitting temperatures. Int J Hydrogen Energy 2018;43:14915–24. https://doi.org/10.1016/j.ijhydene.2018.06.074.
[66] Jacot R, Naik JM, Moré R, Michalsky R, Steinfeld A, Patzke GR. Reactive stability of promising scalable doped ceria materials for thermochemical two-step CO_2 dissociation. J Mater Chem A Mater 2018;6:5807–16. https://doi.org/10.1039/C7TA10966K.
[67] Bhosale RR, Takalkar GD. Nanostructured co-precipitated Ce0.9Ln0.1O2 (Ln = La, Pr, Sm, Nd, Gd, Tb, Dy, or Er) for thermochemical conversion of CO2. Ceram Int 2018;44:16688–97. https://doi.org/10.1016/j.ceramint.2018.06.096.
[68] Bhosale RR, Kumar A, AlMomani F, Alxneit I. Propylene oxide assisted sol–gel synthesis of zinc ferrite nanoparticles for solar fuel production. Ceram Int 2016;42:2431–8. https://doi.org/10.1016/j.ceramint.2015.10.043.
[69] Dey S, Rao CNR. Splitting of CO_2 by Manganite Perovskites to Generate CO by Solar Isothermal Redox Cycling. ACS Energy Lett 2016;1:237–43. https://doi.org/10.1021/acsenergylett.6b00122.
[70] Bhosale RR, Kumar A, AlMomani F, Ghosh U, Khraisheh M. A comparative thermodynamic analysis of samarium and erbium oxide based solar thermochemical water splitting cycles. Int J Hydrogen Energy 2017;42:23416–26. https://doi.org/10.1016/j.ijhydene.2017.03.172.
[71] Padella F, Alvani C, Barbera A La, Ennas G, Liberatore R, Varsano F. Mechanosynthesis and process characterization of nanostructured manganese ferrite. Mater Chem Phys 2005;90:172–7. https://doi.org/10.1016/j.matchemphys.2004.10.033.
[72] Bhosale R, Kumar A, AlMomani F. Solar Thermochemical Hydrogen Production via Terbium Oxide Based Redox Reactions. International Journal of Photoenergy 2016;2016:1–9. https://doi.org/10.1155/2016/9727895.
[73] Chueh WC, Falter C, Abbott M, Scipio D, Furler P, Haile SM, et al. High-Flux Solar-Driven Thermochemical Dissociation of CO_2 and H_2O Using Nonstoichiometric Ceria. Science (1979) 2010;330:1797–801. https://doi.org/10.1126/science.1197834.
[74] Takalkar GD, Bhosale RR, Kumar A, AlMomani F, Khraisheh M, Shakoor RA, et al. Transition metal doped ceria for solar thermochemical fuel production. Solar Energy 2018;172:204–11. https://doi.org/10.1016/j.solener.2018.03.022.
[75] Roeb M, Säck J-P, Rietbrock P, Prahl C, Schreiber H, Neises M, et al. Test operation of a 100kW pilot plant for solar hydrogen production from water on a

solar tower. Solar Energy 2011;85:634–44. https://doi.org/10.1016/j.solener.2010.04.014.
[76] Bhosale RR, Shende R V., Puszynski JA. Sol-Gel Derived $NiFe_2O_4$ Modified with ZrO_2 for Hydrogen Generation from Solar Thermochemical Water-Splitting Reaction. MRS Proceedings 2012;1387:mrsf11-1387-e09-07. https://doi.org/10.1557/opl.2012.772.
[77] Fresno F, Fernández-Saavedra R, Belén Gómez-Mancebo M, Vidal A, Sánchez M, Isabel Rucandio M, et al. Solar hydrogen production by two-step thermochemical cycles: Evaluation of the activity of commercial ferrites. Int J Hydrogen Energy 2009;34:2918–24. https://doi.org/10.1016/j.ijhydene.2009.02.020.

83a Evolving Robust and Interpretable Enzymes for the Bioethanol Industry

Anni Li, Yijie Sheng, Minghui Wang, Xiujuan Li, He Huang

The large-scale production of bioethanol, one of the most promising sustainable renewable resources, holds the potential to address the world's current environmental and energy challenges.[1] Since the widespread production of 1st generation ethanol from sugar- and starch-containing agricultural resources would result in a food supply crisis,[2] integrating lignocellulosic biomass into 1G ethanol processes has become a transition strategy.[3] Lignocellulosic biomass can replace fossil resources if it can be converted into ethanol economically and efficiently.[4] The main components of lignocellulosic biomass are cellulose, hemicellulose, and lignin, which can be chemically or enzymatically hydrolyzed to produce fermentable monomers after appropriate pretreatment.[5] Enzymatic hydrolysis is a practical method for biomass hydrolysis using less energy under mild conditions.[6] Stable cellulases play a critical role in bioethanol fermentation, benefitting from reduced enzyme addition and fewer inhibitor effects.[7] However, a remaining challenge in bioethanol production is that gradually increased ethanol concentrations significantly reduce cellulase activity or even lead to enzyme inactivation.[8]

In our study, we employed endo-β-1,4-glucanase Cel5A from Penicillium verruculosum (PvCel5A) as a model enzyme to investigate the feasibility of applying an evolved robust cellulase in industrial bioethanol production. PvCel5A is promising endoglucanase for lignocellulose hydrolysis due to its high activity.[9] The optimized InSiReP (named InSiReP 2.0) was developed to generate an all-round enzyme, including three steps: (i) eleven potential amino acid positions were obtained by cepPCR; (ii) seven beneficial positions and12 beneficial substitutions were identified by SSM with the 20c-Tangmethod. (iii) InSiReP 2.0 analysis and recombination. The obtained beneficial substitutions were grouped based on the CompassR rule: Subset 1: ΔΔGfold ≤+0.36kcal/mol, predictable beneficial recombination; Subset 2:0.36 < ΔΔGfold ≤+7.52kcalmol 1, unpredictable recombination. Then, the single substitutions of Subset 1 were recombined in silico and ranked according to their thermodynamic stability (ΔΔGfold). The top 10% of ranked recombinants were subjected to gene construction, expression, and experimental validation. The improved recombinants were further recombined with single substitutions in Subset 2. Finally, five all-round PvCel5A variants (V1:S3T/S12K/T77M/E85Y/G264H/D290G; V2:S3T/S12C/T77M/E85Y/G264R; V3:S3T/S12K/T77M/E85Y/G264H; R6:S12K/T77M/E85Y/G264H/D290G; R2:S12K/ T77M/E85Y/G264R/D290G) were yielded with minimal experimental effort. Therefore, the InSiReP 2.0 strategy was proved to successfully engineer PvCel5A with significant ethanol tolerance (up to 7.5-fold improvement in 60 % (v/v) ethanol), thermostability (t1/2 up to 1.42-fold at 75 °C, Tm improved by 3.1 and 5.4 °C in buffer and 40 % (v/v) ethanol, respectively), and tolerance to six representative OSs. The detailed molecular understanding based on MD simulation and the inherent rationality of the InSiReP 2.0 approach (based on the internal stability of enzymes during evolution) enabled us to confidently apply these interpretable cellulases in industrial production processes. Notably, purified PvCel5A R6 (S12K/T77M/E85Y/G264H/D290G) increased the ethanol yield by 8.47 % compared to no cellulase addition in the ADISET process. At the same time, up to 10.08 % improvement of the ethanol yield was obtained when applying PvCel5A R6 in the bioethanol fermentation process of corn stover compared to the addition of only commercial cellulase. The successful application of these robust cellulases, especially PvCel5A R6, in both 1st and 2nd generation ethanol production processes demonstrated the great prospects of using engineered cellulases in industrial production, even beyond the bioethanol field.

Reference

(1) Qiao, J.; Cui, H.; Wang, M.; Fu, X.; Wang, X.; Li, X.; Huang, H. Integrated Biorefinery Approaches for the Industrialization of Cellulosic Ethanol Fuel. Bioresource Technology 2022, 360, 127516. https://doi.org/10.1016/j.biortech.2022.127516.

(2) Schulte, L. A.; Dale, B. E.; Bozzetto, S.; Liebman, M.; Souza, G. M.; Haddad, N.; Richard, T. L.;

Basso, B.; Brown, R. C.; Hilbert, J. A.; Arbuckle, J. G. Meeting Global Challenges with Regenerative Agriculture Producing Food and Energy. Nat Sustain 2022, 5 (5), 384–388. https://doi.org/10.1038/s41893-021-00827-y.

(3) Nair, R. B.; Kalif, M.; Ferreira, J. A.; Taherzadeh, M. J.; Lennartsson, P. R. Mild-Temperature Dilute Acid Pretreatment for Integration of First and Second Generation Ethanol Processes. Bioresource Technology 2017, 245, 145–151. https://doi.org/10.1016/j.biortech.2017.08.125.

(4) Liu, Y.-J.; Li, B.; Feng, Y.; Cui, Q. Consolidated Bio-Saccharification: Leading Lignocellulose Bioconversion into the Real World. Biotechnology Advances 2020, 40, 107535. https://doi.org/10.1016/j.biotechadv.2020.107535.

(5) Han, J.; Luterbacher, J. S.; Alonso, D. M.; Dumesic, J. A.; Maravelias, C. T. A Lignocellulosic Ethanol Strategy via Nonenzymatic Sugar Production: Process Synthesis and Analysis. Bioresource Technology 2015, 182, 258–266. https://doi.org/10.1016/j.biortech.2015.01.135.

(6) Aditiya, H. B.; Mahlia, T. M. I.; Chong, W. T.; Nur, H.; Sebayang, A. H. Second Generation Bioethanol Production: A Critical Review. Renewable and Sustainable Energy Reviews 2016, 66, 631–653. https://doi.org/10.1016/j.rser.2016.07.015.

(7) Teter, S. A.; Sutton, K. B.; Emme, B. 7 - Enzymatic Processes and Enzyme Development in Biorefining. In Advances in Biorefineries; Waldron, K., Ed.; Woodhead Publishing, 2014; pp 199–233. https://doi.org/10.1533/9780857097385.1.199.

(8) Li, X.; Wang, M.; Qiao, J.; Huang, H. Computer-Assisted Enzyme-Cocktail Approach Highly Improves Bioethanol Yield. ACS Sustainable Chem. Eng. 2021, 9 (42), 14277–14287. https://doi.org/10/gpk39b.

(9) Morozova, V. V.; Gusakov, A. V.; Andrianov, R. M.; Pravilnikov, A. G.; Osipov, D. O.; Sinitsyn, A. P. Cellulases of Penicillium Verruculosum. Biotechnology Journal 2010, 5 (8), 871–880. https://doi.org/10.1002/biot.201000050.

Figure. (a) The production performance of corn ethanol fermentation combined with in-situ cellulase-based pretreatment. No cellulase was added to 1; Only crude PvCel5A R6 was added to 2; Only commercial cellulase was added to 3. Equal amounts of commercial cellulase (U/mL) and PvCel5A (U/mL) were added to cellulase cocktails 4-9, which contain different with PvCel5A variants (4: crude PvCel5A WT; 5: crude PvCel5A V1; 6: crude PvCel5A R6; 7: purified PvCel5A WT; 8: purified PvCel5A V1; 9: purified PvCel5A R6). The crude enzyme was concentrated, and all experiments were based on the fermentation of 100 g of liquefied mixture. (b) The ethanol production performance in the fermentation of corn stover with the addition of different cellulases. Equal amounts of commercial cellulase (U/mL) were added to cellulase cocktails 2-5 (2: purified PvCel5A WT; 3: purified PvCel5A V1; 4: purified PvCel5A V2; 5: purified PvCel5A R6). All experiments were based on 10 g of dilute alkali pretreated corn stover. (c) ADISET process. Conventional corn fermentation with SSF (in black) was combined with in-situ pretreatment (in green). (d) The fermentation process of corn stover to yield ethanol through pretreatment, enzyme hydrolysis, and fermentation

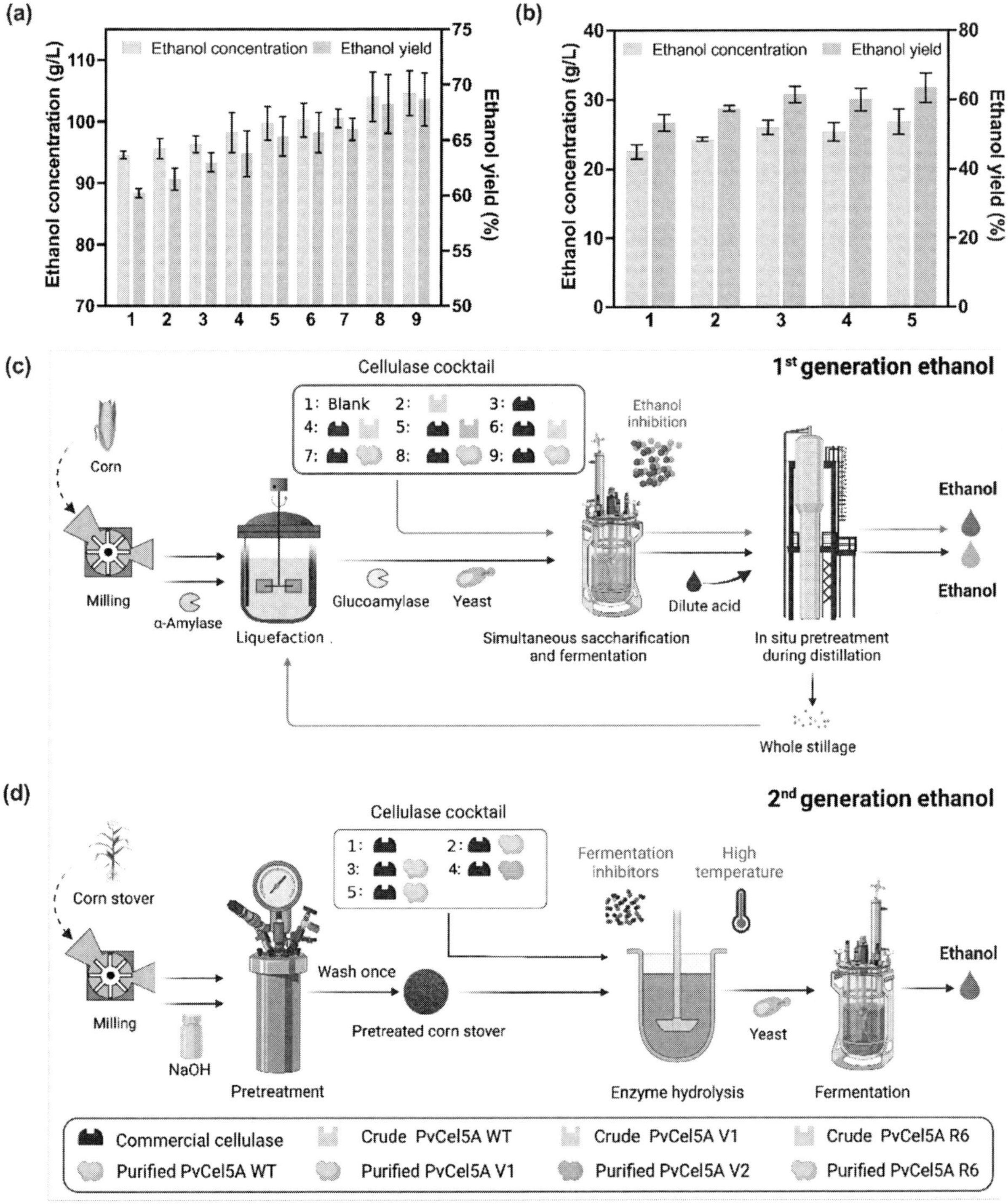
(a)
Ethanol concentration
Ethanol yield
Ethanol concentration (g/L)
Ethanol yield (%)
(b)
Ethanol concentration
Ethanol yield
Ethanol concentration (g/L)
Ethanol yield (%)
(c)
1st generation ethanol
Cellulase cocktail
1: Blank
Ethanol inhibition
Corn
Milling
α-Amylase
Liquefaction
Glucoamylase
Yeast
Simultaneous saccharification and fermentation
Dilute acid
In situ pretreatment during distillation
Whole stillage
Ethanol
Ethanol
(d)
2nd generation ethanol
Cellulase cocktail
Fermentation inhibitors
High temperature
Corn stover
Milling
NaOH
Pretreatment
Wash once
Pretreated corn stover
Yeast
Enzyme hydrolysis
Fermentation
Ethanol
Commercial cellulase
Crude PvCel5A WT
Crude PvCel5A V1
Crude PvCel5A R6
Purified PvCel5A WT
Purified PvCel5A V1
Purified PvCel5A V2
Purified PvCel5A R6

83b In-house Developed Resilient and Sustainable Material for Enhanced Oil Recovery

Syed Muhammad Shakil Hussain, Muhammad Shahzad Kamal, Afeez Gbadamosi, Shirish Patil, Dhafer Al-Shehri

The increase in demand for oil and decrease in supplies requires enhanced oil recovery (EOR) techniques. The waterflooding method of EOR needs the addition of surface-active agents such as surfactants, thus creating the potential for pollution. The produced fluid is a mixture of oil, water, and surfactant which creates pollution of surface waters and land through the leaks or spills of injected chemicals. Therefore, the development of resilient and sustainable surfactants which can bear the harsh reservoir conditions that is stable at 100°C reservoir temperature and soluble in 240,000 ppm salinity of formation water is highly desirable.

To address the environmental and stability concerns, the synthesis of new surfactants was conducted using no solvent or green solvents. The chemical structure of the surfactants was designed in a way to incorporate the cleavable bonds such as amide which have shown good biodegradation, less toxicity, and are environmentally benign. The chemical structure was elucidated by NMR (1H and 13C), FTIR, and mass analysis. The heat tolerance and water solubility tests were done by Thermal Gravimetric Analysis (TGA) and aging techniques. The IFT (interfacial tension) and surface tension were measured using the spinning drop method and force tensiometer (Biolin Scientific), respectively. The rheological parameters were identified by the discovery hybrid rheometer (DHR-3).

The in-house developed surfactants showed excellent solubility in low and high salinity brine and TGA experiments revealed degradation temperature over 100°C which is superior to real oilfield temperature (≥90°C). In addition, the samples were aged in the oven for 3 months at 90°C and no decomposition in chemical structure was observed. The cmc, surface tension at cmc, and IFT of the synthesized surfactants were comparable or higher to commercial surfactants.

According to rheological data, the storage modulus was reduced by increasing the amount of surfactant at a lower rate of shear and frequency because of charge screening and polymer interaction.

The in-house developed resilient and sustainable surfactants demonstrated excellent aqueous and heat stabilities and revealed great potential in harsh conditions of carbonate reservoirs.

83c Assessing the Technical Feasibility of Steam Hydrogasification of Agriculture-Derived Wastes for Renewable Fuel Production in the Leading Agriculture Region

Zhongzhe Liu, Chan Seung Park, Partho Roy, Xin Fan, Marco Ceja, Rosely Ayala, Robert Lozano, Serina Ishida

California is the largest producer of agricultural products in the nation. In particular, the Central Valley of California is one of the world's most productive agricultural regions and many Central Valley's counties are at the top of the nation's agriculture producing counties. Hence, a huge number of agricultural wastes such as corn stover, orchard pruning, and manure is generated regionally every year. However, conventional treatment methods such as open burning and land application lead to air quality issues and contamination of ground water. Additionally, the Central Valley is home to the largest concentration of dairies in California, which has the largest milk producer in the United States. Though industrial wastewater treatment facilities were established for treating regional dairy wastewater, the high strength of dairy wastewater influent is still challenging to traditional treatment processes. Therefore, sustainable methods for processing these agriculture-derived waste streams are highly demanded. Steam hydrogasification reaction based process is an advanced and self-sustainable technology that can convert high moisture content feedstock (e.g., commingled orchard pruning and dairy wastewater) into renewable energy and fuels without costly drying process. In this study, the availability of various agricultural waste streams was assessed, and the technical feasibility of steam hydrogasification of dairy wastewater and other agricultural waste was evaluated experimentally. The preliminary results showed that over 15 million bone dry tons of agricultural wastes are generated in the Central Valley annually, which occupy over 60% of total gross agricultural wastes in California. Also, the dairy wastewater amount is sufficient to be used and mixed with most agricultural waste streams in the Valley. The experimental data showed that steam hydrogasification of dairy wastewater and different agricultural waste led to high carbon conversion (>90%) and desired syngas composition (e.g., sufficient hydrogen for internal recycle use, high methane yield) for further downstream upgrading to final renewable fuel products such as synthetic natural gas.

83d Multiphase Particle in Cell Simulation Study of Ammonia Co-Combustion in a Circulating Fluidized Bed Combustor

Jester Lih Jie Ling, Han Saem Park, Ha Eun Lee, Seung Seok Oh, Hyun Jun Park, Sang Mun Jeong, See Hoon Lee

In this study, a lab-scale circulating fluidized bed (CFB) combustor established by three dimensional Eulerian-Lagrangian model was used to simulate the co-combustion behavior of ammonia and coal. The 3D CFB model consists of riser, cyclone, loop-seal and diplegs. On the basis of the multiphase particle-in-cell (MP-PIC) scheme, the gas phase is modeled by a large eddy simulation (LES) approach and the solid phase by discrete particle method. Various reaction kinetics including devolatilization, combustion of char, volatiles and ammonia, formation and retention of acid gas, and production and reduction of NO and N2O are simultaneously considered. The gas-solid behaviors of the CFB combustor and co-combustion characteristics are obtained with the ammonia co-combustion ratios from 0 to 40%. Numerical results reveal that a fast fluidization regime takes place in the chamber region, while a bubbling fluidization regime appears in the loop-seal. Though N2 input increased with increasing ammonia input, NOx concentration in flue gas was stabilized due to the decrease of thermal NOx formation.

83e Hydrogen Production By Oxidation of Aluminum Nanopowder in Water Under the Action of Laser Pulses

Yaroslav Kraft, Boris Aduev, Denis Nurmukhametov, Gennadiy Belokurov, Natalia Nelubina, Zinfer Ismagilov

Introduction

One of the most promising sources of energy for the future is hydrogen, an environmentally friendly fuel with a high calorific value. Currently, up to 96 % of hydrogen is produced from fossil fuels, and another 4 % from water [1, 2]. Several methods are used for producing hydrogen from water: electrolytic [3], photolytic [4, 5], and combinations of these methods [6, 7]. One of the promising methods for producing hydrogen is based on the chemical interaction of water with metals, such as aluminum or magnesium. In particular, the reaction of one gram of aluminum with water can produce 1.24 liters of hydrogen under normal conditions [8, 9]. The disadvantage of this method is that under normal conditions, metal plates and micron-sized aluminum particles have low activity due to the dense oxide film [10].

The interaction of aluminum nanopowder with water was studied in [11, 12]. The authors of [11] heated an aqueous suspension of aluminum nanoparticles to a temperature of 64–66 °C, which led to an acceleration of the reaction, hydrogen evolution, and water heating. The transformation conditions were found under which the residual content of metallic aluminum in the reaction products reached a minimum value of 1.42 wt %.

In this work, instead of heating, it is proposed to use the action of laser pulses on an aqueous suspension of aluminum nanoparticles in order to achieve complete conversion of aluminum and the maximum rate of hydrogen release.

Materials and methods

The experiments were carried out using aluminum powder prepared by gas-phase synthesis at the Institute of Metal Physics, Ural Branch of the Russian Academy of Sciences. The arithmetical mean size of the particles ranges from 100 to 120 nm. The fraction of metallic aluminum in nanoparticle was equal to (65.0 ± 0.6) wt %.

The experiments were carried out by the following method. 35 mL of distilled water was poured into a glass test tube and 10.5 mg of aluminum nanoparticles was added, which corresponded to a concentration of 0.03 wt %. The test tube was placed in an ultrasonic bath and sonicated for 20 min for uniform distribution of aluminum nanoparticles in the volume. Then the suspension was placed in an experimental 57 mL cuvette, which was sealed. In the initial state the cuvette is not transparent to laser radiation. Sedimentation of nanoparticles becomes noticeable in ca. 3 hours after the start of light transmission in the upper part of the cuvette. Hence, the distribution of nanoparticles over the volume during the experiment can be considered as homogeneous.

The functional diagram of the experimental setup is shown in Fig. 1.

A YAG:Nd^{3+} laser (1064 nm, 14 ns, 10 Hz) was used as a source of laser radiation (1). The energy density was controlled using neutral light filters (2). The laser beam through the lens (3) directed perpendicular to the lateral surface of the cuvette. At a lateral surface of the cuvette the beam diameter was 5 mm. On the opposite surface of the cuvette the beam diameter was 4.6 mm. Thus, in the case of weak absorption, laser energy density in the cuvette is nearly uniform. There are two sealed outlets on the cuvette, which were connected to valves (5) and (6) using a gas sampling line. The gas generated as a result of laser action is directed by means of a valve (6) into a measuring cylinder (7) filled with water, placed in a vessel with water (8), or to a gas analyzer (9). The volume of the formed gas was determined by direct measurement of the water displaced from the measuring cylinder (7) with an accuracy of 0.1 mL. The gas composition was determined using an SRS QMS 300 gas analyzer (9). When determining the gas composition, the air in the cell (4) and gas sampling lines was blown with argon through the valve (5).

Experimental

The experiments were carried out at room temperature $T_0 = 25$ °C. In the absence of radiation, a release of hydrogen from the suspension for 1 hour was not observed.

Under laser irradiation, as a result of the reaction between aluminum and water, water separation occurs with the formation of hydrogen. The dependence of the increase in the volume of gas in the measuring cylinder on the time of irradiation of the suspension by laser pulses with a frequency of 10 Hz was measured. The radiation energy density used was 0.5; 1.0; 2.0; 3.0 and 6.0 J/cm^2. The energy of laser pulses on a spot 5 mm in diameter was varied within 0.1–1.18 J. In each experiment, we used the same weighed amount of aluminum nanoparticles in water equal to 10.5 mg. The result is shown in Fig. 2.

The maximum volume (V_m) of the produced gas was the same for all laser energy densities used and amounted to V = 9 mL. In this case, the time to reach the maximum value depends on the laser radiation energy density H, varying from 60 min at H = 0.5 J/cm^2 to 3.5 min at H = 6.0 J/cm^2. The suspension in the cell, which had initially gray color, during irradiation was bleached and became transparent at attaining the V_m.

The self-heating temperature of the total mass of the suspension during the reaction did not exceed 15 °C.

Experimental points in Fig. 2 are approximated by the formula

$V = V_m [1 - \exp(-t/\tau)]$ (1)

The characteristic times τ of increasing the gas volume to $0.63V_m$ under the exposure of the suspension at a certain laser energy density are listed in Table 1; correlation coefficients R^2 are also listed there. Thus, dependences of the volume of nascent gaseous products on the exposure time of suspension at the fixed energy densities are satisfactorily described by formula (1).

With the gas analyzer SRS QMS 300, the composition of the gaseous products of the interaction of aluminum nanoparticles with water was studied after the pulsed-periodic laser exposure for 5 minutes with an energy density of 2.0 J/cm^2. The result is shown in Table 2.

A control experiment was performed. Water without aluminum nanoparticles was irradiated under similar conditions. Within the measurement error of a gas analyzer, oxygen and hydrogen were not detected. Hence, the hydrolysis of water is absent.

Tables 1 show the characteristic times needed to decrease the concentration of metallic aluminum τ in the suspension by a factor of 2.72 depending on the laser radiation density H. The τ values were used to plot the dependence $1/\tau = f(H)$. The result described by the linear dependence is displayed in Fig. 3. Hence, the characteristic time of metallic aluminum conversion in nanoparticles can be described by the hyperbola formula:

$\tau = \alpha/H$ (2)

where $\alpha = 4.5$ (min·J/cm^2) was estimated from the slope of the line in Fig. 3. The α value indicates the characteristic time required for decreasing the concentration of metallic aluminum by a factor of 2.72 under the action of a unit laser energy density.

Conclusions

It was shown that hydrogen can be produced under the action of laser pulses on the suspension containing 0.03 wt % of aluminum nanoparticles in water (10.5 mg of aluminum in 35 ml of water). Aluminum nanoparticles were found to absorb the energy of laser radiation. The maximum values of the produced hydrogen volume in the laser energy density range of 0.5–6 J/cm^2 are close and amount to 8.4 mL. The time of complete transformation of metallic aluminum decreases with an increase in the energy density of laser radiation according to the hyperbolic law and reaches 3.5 min at $H = 6\ J/cm^2$.

References

1. Wang HZ, Leung DYC, Leung MKH, et al. Renew Sust Energ Rev 2009;13:845–53.
2. Balat M. Int J Hydrogen Energy 2008;33:4013–29.
3. Kaya MF, Demir N, Albawabiji MS, et al. Int J Hydrogen Energy 2017;42:17583–92.
4. Juodkazytė J, Seniutinas G, Šebeka B, et al. Int J Hydrogen Energy 2016;41:11941–8.
5. Huang C, Linkous CA, Adebiyi O, et al. Environ Sci Technol 2010;44:5283–8.
6. Geraldo de Melo Furtado J, Gatti GC, Serra ET, et al. Int J Hydrogen Energy 2010;35:9990–5.
7. Chang ACC, Chang H-F, Lin F-J, et al. Int J Hydrogen Energy 2011;36:14252–60.
8. Shevchenko VG, Kononenko VI, Lukin NV, et al. Combust Explos Shock Waves 1999;35:77–9.
9. Tsuchida T, Hasegawa T, Kitagawa T, et al. J Eur Ceram Soc 1997;17:1793–5.
10. Pityulin AN, Shcherbakov VA, Borovinskaya IP, et al. Combust Explos Shock Waves 1979;15:432–7.
11. Godymchuk AYu, Ilyin AP, Astankova AP. Bulletin of the Tomsk Polytechnic University 2007;310:95–7.
12. Ilyin AP, Korshunov AV, Tolbanova LO. Bulletin of the Tomsk Polytechnic University 2007;311:10–4.

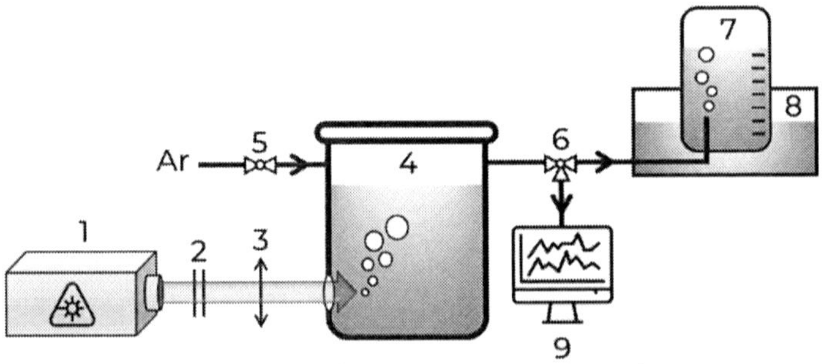

Fig. 1 – Functional diagram of the experimental setup

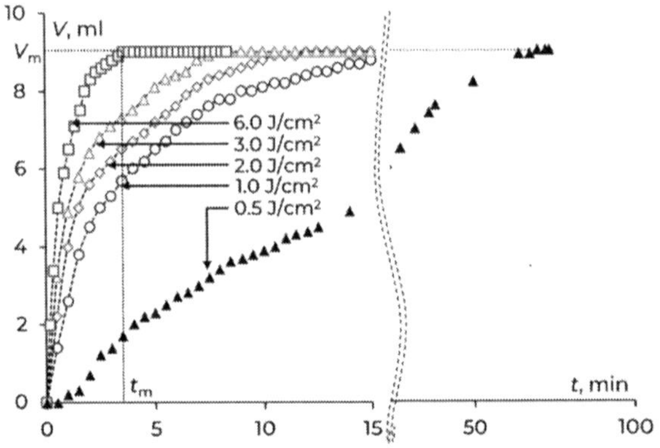

Fig. 2. – The gas release on the time dependence under the irradiation of 0.03 % Al suspension in water by the laser pulse (1064 nm, 14 ns, 10 Hz) with different energy densities; V_m is the maximum gas volume

Table 1 – Characteristic times τ corresponding to an increase in the volume of released gas to $0.63V_m$ at respective laser energy densities H

H	τ, min	R^2
0.5	18.0	0.90
1	3.7	0.92
2	2.5	0.93
3	1.2	0.92
6	0.8	0.96

where R^2 is the correlation coefficient for approximation of experimental points in Fig. 2 by formula (1)

Table 2 – Component composition of gaseous products from the interaction of aluminum with water after repetitively pulsed laser exposure

Components	H_2	H_2O	N_2	O_2
Volume fraction, %	93.8 ± 0.1	3.1 ± 0.2	2.5 ± 0.2	0.6 ± 0.1

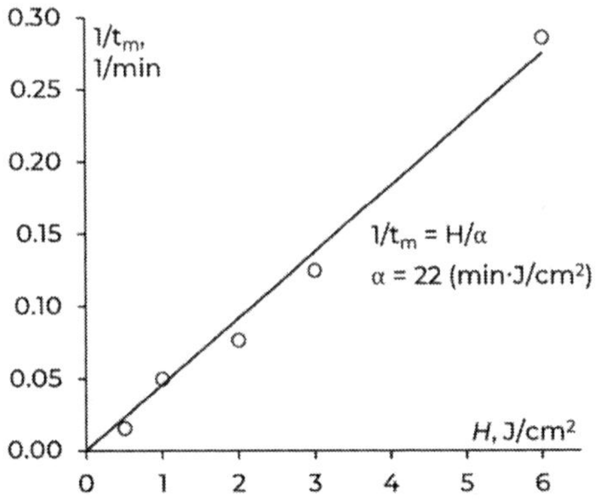

Fig. 3. – Dependence of the inverse time of complete oxidation of aluminum nanoparticles on the laser energy density.

83f Performance Targets for Oxidative Coupling of Methane from Techno-Economic Profiling

Faisal Ashour, Mamoun Al-Rawashdeh, Patrick Linke

The Oxidative Coupling of Methane (OCM) route to convert methane to ethane and ethylene (C2) has been investigated as a potentially promising path for natural gas monetization for more than 40 years. Significant research efforts have been invested in the field of catalysis accompanied by considerably less efforts in process design and analysis to establish economic performance. The current state-of-the-art in catalysis and reaction engineering still results in drastic degradations of C2 selectivity with increases in conversion. This poses challenges for economic process performance as value addition to the methane feed through conversion to C2 products, which is generally bounded by the price difference between reactants and products, is offset by the high costs associated with low conversion.

This work introduces a systematic method to explore the minimum performance requirements for OCM reactors that could enable commercial viability through techno-economic profiling. The multi-level methodology aims to determine the minimum requirements for the key reactor level performance indicators selectivity and conversion based on high-level economic analysis. At the initial level, high-level value added calculations considering raw material, product and energy prices as well as carbon taxes at the input-output level of an OCM process are performed to establish minimum selectivity requirements and visualize the OCM process in the form of 'value addition/destruction maps' that shows revenues, costs and gross value added as functions of C2 selectivity for ranges of ethylene to ethane selectivity. This analysis sets an absolute minimum requirement for reactor selectivities that must be achieved at full conversion within a no capital cost OCM process. The presented work will show how most published studies already remain below the absolute minimum selectivity requirement and will lead to value destruction. At the next level, major additional costs related to separations and recycling form incomplete conversions are included to set a corresponding lower bounds for both selectivity and reactor conversions that must always be exceeded by a potentially profitable process. Minimum energy requirements for separations and recycling are calculated for complete ranges of selectivity and conversions to determine the conversion—selectivity frontier that bounds the value destruction domain. Analysis shows that reported OCM reactor performances are exclusively located within the value destruction regime. Further levels of analysis explore sensitivities of the selectivity-conversion frontier to deviations from the idealized performance assumptions towards efficiencies and costs of real systems to generate aspirational targets be exceed by future efforts in catalysis and reaction engineering to achieve profitable OCM.

While the presented method of techno-economic profiling is illustrated for the OCM route, it is generally applicable to the study of any reaction route. The quick analysis enables an early and clear understanding of process economics and reactor level performance bounds that can be used to steer time and resource intensive development efforts in catalysis and reaction engineering from a process perspective.

529a Creating a Ternary Phase Diagram for Pyrene and Its Oligomers: The Effect of Oligomeric Composition on Liquid Crystalline Phase Behavior

Jefferson Giles II, Mark C. Thies

More than 50 years ago, Brooks and Taylor (B&T) at Union Carbide discovered that petroleum pitch forms a liquid crystalline phase, or mesophase, after proper annealing. This mesophase is a necessary precursor for graphite formation in petroleum pitch. Today, specialized mesophase pitches derived from petroleum can be graphitized to make carbon fibers with exceptional tensile modulus and thermal conductivity.

Ironically, since B&T's discovery in 1965, our fundamental understanding of the role that molecular size, molecular structure, and mixture composition play in the properties of both the pitch precursor and the final carbon products continues to be quite limited. For example, the ideal molecular size for producing a mesophase pitch with a low melting point, a highly desirable property for processing, is unknown, as are the most desirable mixture compositions for forming low-melting mesophases.

Even with the recent significant advances in molecular characterization, the literally thousands of molecules present in petroleum pitches are a significant barrier to the development of structure–property relationships for pitches. To make the connection between molecular structure and final properties, we need a pitch system consisting of just a few components – with some of those components being isotropic and some being mesogenic.

We therefore propose the use of non-fragmenting pyrene oligomers, which are known to consist of both isotropic and mesogenic components. Although our group discovered that trimer forms 100% mesophase, here our interests lie with mixtures of these oligomers and their mesogeneity. To help us create these mixtures, we are using our unique supercritical extraction (SCE) column setup to isolate both pure pyrene oligomers and well-defined binary and ternary mixtures of said oligomers. This information will be used to generate one of the first true ternary phase diagrams of isotropic–mesophase behavior for a model system based on petroleum pitches.

529c Development of a Robust Thermodynamic Model to Predict the Onset Crystallization Temperature of Fatty Acid Methyl Esters(FAMEs)

Junli Liu, Bernard Tao

Current biodiesel mainly consists of fatty acid methyl esters and is viewed as an alternative to petroleum diesel for renewability and sustainability. Because the fatty acid profiles significantly varied for the sources, the compositions significantly affected the biodiesel properties, such as low-temperature performance. The cloud point of biodiesel indicates the phenomenon of solid-liquid phase change and can characterize the low-temperature performance. The statistical and phase-equilibrium models have been developed to predict the cloud point. However, these models for cloud point predictions were limited to known components in the mixtures. The binary and ternary mixtures of FAMEs were prepared from pure components, while multi-component mixtures were obtained from biodiesel with various feedstocks and fractionated with urea inclusion. Cloud points of all these mixtures were measured according to ASTM D 2500. With the application of phase equilibrium principles, a thermodynamic model was proposed, and the non-ideal behaviors were characterized by activity coefficients which were predicted by the UNIFAC model and Modified UNIFAC models. The developed thermodynamic model accurately predicts the cloud point according to compositions. The model extends the application scope to the low-temperature range. There had eutectic points in the phase diagrams when the mixtures consisted of either saturated FAMEs or unsaturated FAMEs. This study proved that saturated FAMEs regulate the cloud points, but unsaturated FAMEs affect them through group interactions. Moreover, the proposed model can predict unknown FAMEs mixture as it is built on the group contributions.

529d Predictive Parametrization of PC-SAFT EoS By Neural Network Ensembles Based on Molecular Fingerprints

Jonas Habicht, Christoph Brandenbusch, Gabriele Sadowski

Simulation of chemical or biotechnological processes plays a key role in the design or optimization of industrial processes or production plants. Especially in the design of modern biofuels, physical properties of a variety of different hydrocarbons (ethers, oxygenated hydrocarbons, fatty acid methyl esters) are needed, as the development of new feedstocks for fuel production are a significant challenge of our time. For this purpose, the phase behavior can be efficiently described using advanced equations of state (EoS), such as PC-SAFT. These equations of state (as most thermodynamic models) require a parametrization (pure-component parameters) of each component. This is typically performed by fitting the respective parameters to physical pure-component properties (e.g. vapor pressure, saturated liquid density). However, such experimental data is often not available, especially if complex molecules such as pharmaceuticals are considered. Thus, parametrization using non-experimental inputs are of high interest.

In this work, a neural network (NN)-ensemble was developed to predict PC-SAFT pure-component parameters for non-associating molecules. Beside basic molecular input features (e.g. molar mass, number of rotatable bonds), extended-connectivity fingerprints (ECFPs) were used as key input feature to characterize the molecules. All input features were directly derived from easy available SMILES-notation of the molecules. A dataset comprising ~300 molecules was used for training of the NN-esemble. To increase the statistical validity, 5-fold cross validation with three random initial splits was performed creating a NN-ensemble of 15 NNs to predict the three PC-SAFT pure-component parameters for non-associating molecules. The ensemble prediction yielded good accuracy (AARD < 8 % between ML-predicted PC-SAFT parameters and literature PC-SAFT parameters) for the considered test dataset. Moreover, ML-predicted PC-SAFT pure-component parameters for unknown molecules (not included in training) showed a remarkable performance in describing the experimental (validation) data.

This novel ML-approach offers a reliable and easy access to PC-SAFT pure-component parameters of any non-associating molecule solely based on the chemical structure formulated as SMILES string. Prospectively, ML-predicted pure-component parameters can be used in an early state of process design to estimate phase behavior using a physics-based EoS without additional experimental effort.

529e Heat Capacity of Saturated and Compressed Liquid Dimethyl Ether at Temperatures from (132 to 345) K and at Pressures to 35 Mpa: Measurements and Derived Equation of State

Joseph Magee, Jiangtao Wu

Molar heat capacities at constant volume (C_V) of dimethyl ether have been measured with an adiabatic calorimeter [1]. Temperatures range from the triple point to 345 K, and pressures up to 35 MPa. Measurements were conducted on liquid in equilibrium with its vapor and on compressed liquid samples. The samples are of high purity, as verified by chemical analysis. Calorimetric quantities are reported for the two-phase ($C_V^{(2)}$), saturated-liquid (C_σ or C_X'), and single-phase (C_V) molar heat capacities. Low temperature $C_V^{(2)}$ data were employed to estimate vapor pressures for values less than 100 kPa by applying a thermodynamic relationship between the two-phase internal energy $U^{(2)}$ and the temperature derivatives of the vapor pressure. Vapor pressures were calculated at temperatures as low as the triple-point temperature. The principal sources of uncertainty are the temperature rise measurement and the change-of-volume work adjustment. The expanded relative uncertainty (with a coverage factor k=2 and thus a two-standard uncertainty estimate) is estimated to be 0.7 % for C_V , 0.5 % for $C_V^{(2)}$ and 0.7 % for C_σ . NIST's *ThermoData Engine* [2] software (TDE) was used to compare the new measurements with published data. Comparisons were facilitated by a simple 12-coefficient equation of state that was derived with TDE's multiproperty-fitting method by combining thermodynamic properties measurements for heat capacity, density, vapor pressure, speed of sound, plus their standard uncertainties. The derived equation of state could be useful to practicing engineers for industrial process calculations.

[1] Wu, J. and Magee, J. W., 'Heat Capacity of Saturated and Compressed Liquid Dimethyl Ether at Temperatures from (132 to 345) K and at Pressures to 35 MPa' J. Chem. Eng. Data 63: 1713-1723 (2018). [DOI: https://doi.org/10.1021/acs.jced.8b00037]

[2].Diky, V., Chirico, R. D., Frenkel, M., Bazyleva, A., Magee, J. W., Paulechka, E., Kazakov, A. F., Lemmon, E. W., Muzny, C. D., Smolyanitsky, A. Y., Townsend, S. A. and Kroenlein, K., ThermoData Engine (TDE) Version 10.4.4 (Pure compounds, Binary Mixtures, Ternary Mixtures and Chemical Reactions); Standard Reference Database 103b, 'NIST ThermoData Engine, NIST Standard Reference Database 103b-Pure Compounds, Binary Mixtures, Ternary Mixtures and Chemical Reactions, Version 10.4.4', National Institute of Standards and Technology, Gaithersburg, MD (2022). URL https://www.nist.gov/mml/acmd/trc/thermodata-engine/srd-nist-tde-103b

582a Competitive Adsorption and Displacement Behavior of CO_2/CH_4 Binary Mixture in Shale Matrix with Heterogeneous Surfaces

Jingyue Sun, Cong Chen, Zhehui Jin, Weizhong Li

The CO_2 sequestration and enhanced gas recovery (CS-EGR) technology provides a very effective way to alleviate the greenhouse effect and energy crisis. The characteristics and mechanisms of CO_2/CH_4 competitive adsorption and displacement behavior in shale matrix as well as the effects of reservoir parameters play a vital role in enhancing shale gas production and CO_2 storage. However, it has not been fully understood especially in pores with heterogeneous surfaces composed of organic matter and inorganic compounds. A graphene-MMT heterogeneous surface was proposed as a pore model and molecular dynamics (MD) simulations have been applied.

Absolute adsorption selectivity based on the fixed mole fractions of CO_2 and CH_4 in the total shale pores has been proposed to evaluate the competitive adsorption features with relative adsorption selectivity. A strong asymmetric competitive adsorption behavior of CO_2 and CH_4 has been observed. In the case of certain temperature and pore size, the partial pressures of CO_2 and CH_4 are found to play the decisive role in the competitive adsorption by comparing the pores with homogeneous and heterogeneous surfaces. The competitive adsorption behaviors of CO_2/CH_4 near the corresponding surface in two types of pores are consistent when the partial pressure of the components of the mixture in pores with homogeneous surfaces are correspondingly equal to that in pores with heterogeneous surfaces. Therefore, it's better to describe the competitive adsorption behavior by gas partial pressure. Langmuir equation was applied to fitting adsorption isotherm in the first adsorption layer and the second adsorption layer. The good fitting results indicate that although the competitive adsorption mechanism of CO_2/CH_4 mixture can't be considered as monolayer adsorption due to the existence of multiple adsorption layers, each adsorption layer can be explained by Langmuir adsorption theory. To enhance the CH_4 recovery and CO_2 storage, the appropriate timing and amount of CO_2 injection were proposed. First reduce the reservoir pressure to as low as possible by, for example, fracturing, because the first pressure drawdown is obviously beneficial to CH_4 production. Second, inject a certain amount of CO_2 and soak. Third, reduce the pressure to the final pressure of the first pressure drawdown process. It should be noted that the Ratio should be controlled not to exceed 1.11 ~ 1.21 when injecting CO_2 in the second step, otherwise a huge separation cost will be required. After that, much CO_2 should be injected to restore reservoir pressure.

In addition, a CO_2–CH_4 displacement model that can provide a continuous pressure gradient in the displacement direction and a stable back-pressure at the pore outlet was developed based on the graphene-MMT heterogeneous surface pore. The model was divided into three areas: CO_2 injection area, pore area, and backpressure area. It was found that the displacement process always starts from the CH_4 reverse flow stage and then experiences the injection pressure action stage and positive displacement stage in sequence. Moreover, the extent of CH_4 reverse flow directly affects the system development process and the final displacement efficiency. A small system presorption pressure and a large injection pressure are beneficial to the displacement. It is believed that the reservoir pressure should be dropped to the lowest possible level during depressurization exploitation, and the CO_2 injection pressure needs to be selected by considering displacement efficiency, reservoir safety, and economic cost. The CO_2 occupies the adsorption sites near graphene faster than that near montmorillonite (MMT) in the direction of displacement, while the CH_4 desorption is faster near MMT. Therefore, it cannot be concluded that the displacement process near graphene is ahead of MMT. It is considered that the gas desorption/adsorption behavior near the graphene dominates the displacement process in terms of gas amount, while the fluctuation near MMT with time and injection pressure directly affects the gas adsorption variation in the pore space from the trend.

As carbon emissions cause more and more serious damage to the environment, CO_2 geological storage is one of the most concerned solutions. The decisive role of gas partial pressure and the adsorption/displacement mechanism of binary mixture discovered in this study may help to understand the adsorption behavior from a micro perspective, and the operation recommendations can effectively enhance the shale gas recovery and CO_2 storage. This is of great significance to environmental production and alleviation of energy crisis.

582c Integrating Public and Private Data for Modeling and Optimization of Shale Oil and Gas Production

Vyacheslav Romanov, Derek M. Vikara, Kolawole Bello, Shahab D. Mohaghegh, Guoxiang Liu, Luciane Cunha

The overarching goal of this project was to develop novel methods to facilitate rapid data transformation into knowledge, for subsurface applications. In the first phase of this effort, the project team developed workflows and interactive applications that unite data from a variety of public and private sources in machine-readable data formats to drive machine learning (ML) and artificial intelligence (AI) applications, including methods to assess data quality, identify data gaps, fill data gaps through science-based approaches, and demonstrate that science-informed AI/ML can be used for near real-time scenario analysis and optimization of field development planning and resource recovery. The study implemented a data-intensive supervised machine learning approach, through a quasi-experimental framework, with the objective of quantifying the impact of oil and gas operator-specific proprietary data on ML-based predictive model performance as compared to using oil and gas datasets that are publicly available. Using the time-series data, the long short-term memory (LSTM) based neural network models were designed and trained to jointly predict daily oil, gas, and water production for horizontal wells as a function of bottom-hole pressure drawdown, spatial placement across the study domain, and well completion attributes. The combination of input features was set to align with either one of three categories: 1) predominantly proprietary in origin (proprietary model); 2) a mix of data of proprietary and public origin (quasi-public model); or 3) only the data commonly available from public or vendor sources (public model). The best-performing model utilized the proprietary data for downhole pump inlet pressure as a key input attribute to simulate the hydrocarbons production response to various pressure management scenarios. The study specifically examined utility of the proprietary deep learning model developed by the team in forecasting unconventional oil and gas production by using well data from the Permian Basin (Midland sub-basin, Texas, United States). The model predicted daily production for oil, water, and gas with accuracy on the order of 79 percent (for water and gas) to 86 percent (for oil). The extensive input parameter set used in the development of the model provides the utility to test and evaluate multiple controlling features on associated production. These features include (but are not limited to) geologic properties from well log data, artificial lift design data, well placement (spatial, depth, and wellbore trajectory orientation) and completion attributes, detailed well hydraulic fracturing data, well operating conditions, and operational controls of production intensity (via pumping pressure downhole). Results of the scenario planning exercise indicated that rapid drawdown of pressure generated the initially higher oil and gas production. However, the cumulative volumes of oil and gas production in response to rapid drawdown strategies were lower compared to conservative drawdown strategies that sustained pressure over the medium-to-long term. The rapid drawdown strategies also resulted in lower water production over the lifespan of the well compared to conservative drawdown strategies. This production forecast of the varying drawdown strategies may have significant operational and economic implications, with contrasting perspectives between well productivity and profitability given the typical oil and gas economics and the volatility in the oil and gas markets. Further analysis using the proprietary 'shut-in' and 'frac-hit' data provided the means to estimate the potential oil production lost due to 'frac-hit' events at 287 wells in the proprietary dataset made available by the project partners. The historical oil price ($/bbl) was correlated with the dates of 'shut-in' and 'frac-hit' periods. For the entire field, the cumulative sum of all 'lost-production' events returned an estimated annualized loss of approximately $30 million.

591a Direct Conversion of Methane on Metal-Oxide and Oxide-Oxide Interfaces: Fundamental Studies on the Production of Hydrogen and Methanol

Jose Rodriguez

The transformation of methane into H_2 or methanol at moderate temperature and pressure conditions is of great environmental interest and remains a challenge despite many efforts. We have found that on Ni/CeO_2 and Pt/CeO_2 systems strong metal-support interactions modify the chemical activity of the admetals and help to dissociate methane at room temperature. These systems are very efficient for the production of H_2 through the dry reforming of CH_4 with CO_2. Recent studies have shown that several oxide-oxide interfaces (X-CuO; X= ZnO, CeO_2, or SnO) are active for a direct CH_4 to CH_3OH conversion. The selectivity towards methanol production is usually enhanced when water is added to mixtures of methane and O_2. The results of ambient pressure X-ray photoemission spectroscopy and density functional theory calculations have shown that water preferentially dissociates on active sites of the metal-oxide and oxide-oxide interfaces. The adsorbed OH blocks O-O bond cleavage that would dehydrogenate CH_3O to CO and CO_2, and it directly converts the CH_3 species to methanol. Water adsorption also helps to displace the produced methanol into the gas phase. The combination of these three phenomena helps to improve the selectivity for a direct CH_4 to CH_3OH conversion.

591b Rich CH_4 Oxidation on Pt/Pd/Al_2O_3 Monolith: Performance, Multiplicity, and Spatial Effects

Jonathan Ratcliff, Michael Harold

The increased interest in natural gas as a fuel requires improved oxidation catalysts to minimize the slip of methane, a potent greenhouse gas. Under rich conditions ($O_2/CH_4 < 2$) CH_4 is converted to a mixture of total oxidation products (CO_2, H_2O) and partial oxidation products (CO, H_2). We describe the results of an experimental study of rich methane oxidation on Pt/Pd/Al_2O_3 washcoated monolith.

This monolith was introduced to feeds containing a range of different compositions (O_2, CO_2, H_2O) and temperatures. To assess the behaviors of this system, spatial measurement techniques were utilized. Spatial temperature was recorded using optical frequency domain reflectometry (OFDR) and concentration by capillary inlet mass spectrometry (SpaciMS). Figure 1 shows the impact of both variances in inlet O_2 and H_2O. Spatial concentration profiles for a dry feed in the O_2-starved regime (< 2000 ppm) reveals that complete oxidation occurs upstream while anaerobic chemistry such as steam reforming occurs downstream. Increases in O_2 result in initial increases in conversion across all conditions followed by eventual decline in conversion to an oxygen-insensitive value. This decrease occurs due to inhibition by site blockage by oxygen adatoms. Increasing the amount of H_2O present results in increased methane conversion. Figure 2 shows spatial methane conversion profiles for a condition corresponding to the region of declining methane conversions. This profile shows the existence of two separate steady states, one inhibited and one active. The combination of these profiles produces the bulk conversion previously observed with the decline occurring due to more channels being inhibited as opposed to active. A kinetic model to describe this system is being worked on and results will be presented and discussed.

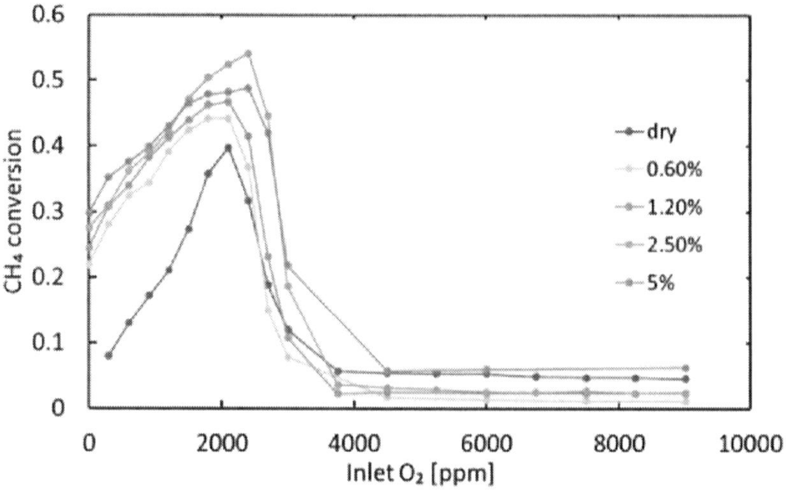

Figure 1: Impact of different inlet H_2O concentrations on CH_4 conversion for 500 °C

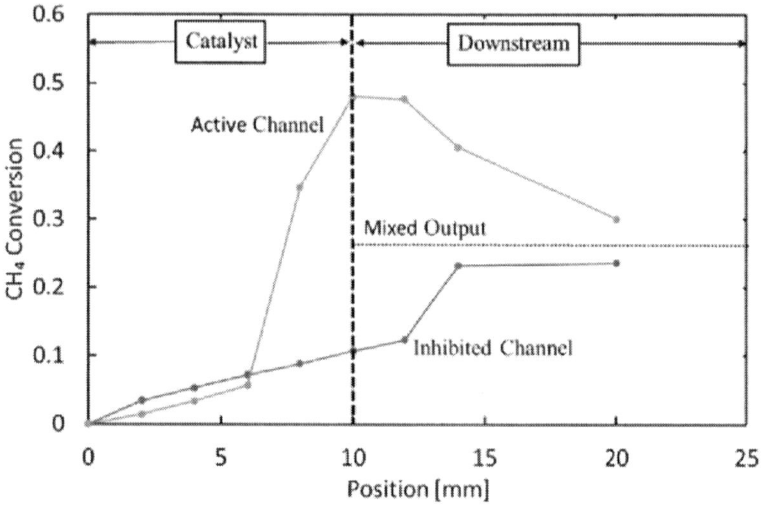

Figure 2: Coexisting steady states for inlet conditions of 3900 ppm O_2, 4200 ppm CH_4, and 525 °C (dry)

591c Aiding Catalysis with Additive Manufacturing: Fabrication and Optimization of Hierarchical Membrane/Catalyst Systems for Oxidative Coupling of Methane

James Wortman, Valentina Omoze Igenegbai, Ali Hussain Motagamwala, Rawan Almallahi, Suljo Linic

One of the largest detriments to C_{2+} hydrocarbon selectivity in the oxidative coupling of methane (OCM) is the undesired homogeneous oxidation of the C_{2+} products with O_2 gas. Therefore, membrane/catalyst systems (which feed O^{2-} species) can potentially allow higher C_{2+} selectivity than equivalent packed bed reactors (which feed O_2(g)). However, parasitic undesired O_2(g) evolution on the OCM side of the membrane can nullify much of the selectivity enhancement. Furthermore, issues of insufficient C_{2+} yields, low reaction rates, and high membrane capital costs have also hindered the development of these membrane/catalyst systems. Here, we present a streamlined, tunable synthesis approach for the design of catalytic hollow fiber membrane OCM systems, inspired by additive manufacturing. The presented approach allows us to carefully co-design membrane and catalyst functionalities with tunable layer porosity and tunable thickness, helping to address the above challenges. As an OCM case study, we use $BaCe_{0.8}Gd_{0.2}O_{3-\delta}$ (BCG) as the basis of both the catalyst and membrane layers. Due to the tuned BCG catalyst surface area for methane activation on the OCM side, the C_{2+} selectivity, yield, and reaction rates were greatly improved compared to a thick, symmetric BCG hollow fiber with non-tunable catalytic properties. A maximum C_{2+} yield of 22.7% was achieved at 845 °C using the optimized system. The enhanced performance of the asymmetric system is attributed to an increased heterogeneous CH_4 activation rate that stoichiometrically matches the incoming O_2 flux, thereby minimizing O_2(g) effects. We use OCM kinetic studies and dimensional analysis to argue that this "rate matching" of O_2 flux and CH_4 activation rate is critical to maximizing the C_{2+} yield, and that this requires careful co-designing of membrane and catalyst functionalities.

591d *Chemical Looping for the Production of Value-Added Oxygenates from CO_2 and Natural Gas*

Sherafghan Iftikhar, Luke Neal, Junchen Liu, Kyle Vogt-Lowell, Vasudev Haribal, Andrew Tong, Fanxing Li

Natural gas is an abundant and relatively low-carbon energy source, but it still contributes to emissions of greenhouse gases. However, when used to make value-added oxygenates such as acetic acid, ethylene glycol, and propionic acid, natural gas's chemical composition is conducive to the beneficial utilization of CO_2. For example, the net reaction of methane with CO_2 to produce acetic acid is mildly exothermic and consumes one mol of CO_2 per mole of methane. Unfortunately, conventional routes to producing acetic acid from natural gas include several parasitic energy losses that prevent the system from achieving net CO_2 consumption. In these routes, natural gas is first converted to syngas in an energy-intensive reforming step, followed by extensive water gas shift and/or cryogenic separation to achieve proper ratios of hydrogen and CO for the needed chemicals. The high temperature of reforming needed for high yields is not conducive to sustainable energy sources such as renewable or waste energy, and the parasitic losses generally require additional energy imports into the system and create a significant net carbon footprint for the process.

To address these issues, we propose to utilize a chemical looping reforming approach. In the first step, methane is partially oxidized with a mixed metal oxide to form a 2:1 H_2 to CO syngas mixture. In a subsequent step, the metal-oxide is regenerated. Although regenerating the metal oxide with air makes the system thermally sufficient, it can also be regenerated with CO_2 to form carbon monoxide. The operating temperature of the process allows the integration of renewable and/or industrial waste heat sources, allowing for the net reaction endotherm to be made up without contributing to the carbon footprint of the process. The inherently separated syngas and CO streams are ideal for efficient chemical synthesis. The syngas can be converted directly to methanol, which subsequently reacts with CO to form acetic acid through carbonylation.

We present experimental and modeling data showing that the chemical looping approach can facilitate the production of acetic acid from natural gas with a negative net CO_2 emission. We also briefly describe chemical looping routes to other value-added chemicals with net utilization of CO_2.

591e Supported Catalytically Active Liquid Metals Solutions (SCALMS) As Catalysts for Ethane Dehydrogenation

Aime Laurent Twizerimana, Gia Monte, Aaditya Hari Bharanidharan, Kaylen Ocampo, Kyle McGaughy, Mohammad S. Masnadi, Goetz Veser

The increase in global ethylene demand combined with the availability of cheap shale gas in the U.S is attracting renewed research interest in catalytic ethane dehydrogenation to ethylene (EDH). Currently, in the U.S, ethylene is produced via steam cracking of ethane, where ethane is mixed with steam and heated to ~ 800 where it dehydrogenates to ethylene in a non-catalytic gas phase reaction. Although this process is widely used commercially, it suffers from multiple issues, including heavy coking (i.e. carbon deposition on the reactor walls) and high CO_2 emissions due to the required high reaction temperature. Use of a catalyst could help to improve selectivity toward ethylene and lower the reaction temperature, hence reducing the emissions associated with heating. However, all catalysts studied to-date for this reaction suffer from rapid deactivation due to coke deposition and often sinter at the high-temperature operation. There is hence a need for designing an active and stable catalyst that can endure high-temperature operation and coke formation.

Liquid metals, i.e. metals with a low melting point, have recently emerged as an interesting novel reaction medium. Currently, liquid metal catalysts can be categorized into two systems: first, liquid metal bubble column reactors (LMBR) that rely on bubbling gaseous reactants through a column of liquid metals. Second, supported catalytically active liquid metal solutions (SCALMS), which mimic conventional supported metal catalysts, i.e. which use conventional solid supports like alumina or silica to support small droplets of the liquid metal. LMBR system combine high heat capacity with excellent resistance to coking by separating the formed coke from the catalytically active molten metal due to density differences. While these properties make LMBR an attractive, robust reaction medium, LMBR require the use of large amounts of molten metals and are challenged by the lack of prior experience with molten metal reactors. Furthermore, LMBR are difficult to characterize using common catalyst characterization techniques since liquid metals are opaque to most regions of the electromagnetic spectrum that are used in conventional spectroscopic techniques.

In contrast, SCALMS offer an opportunity to use and study liquid metals under more easily accessible conditions as they bridge between conventional heterogeneous catalysts and LM systems. They hence enable integrating liquid metals into a conventional heterogeneous catalysis infrastructure, opening a window towards developing a fundamental understanding of how liquid metal catalyst interact with gaseous reactants using conventional solid catalysts characterization techniques. Furthermore, SCALMS provide more efficient use of metals by maximizing the available interfacial area. Most significantly, a few prior reports on the use of SCALMS, focused on Ga-based catalyst systems, reported unexpected resistance to coking in propane and butane dehydrogenation, but did not identify the mechanism by which SCALMS resist deactivation.

In the present work, we are exploring Bi-based SCALMS for ethane dehydrogenation in order to verify the presence of this unusual coking resistance and explore its origins. A particular focus of the studies is the synthesis of SCALMS with good control of particle size and composition in order to obtain the well-defined catalyst systems needed to understand the fundamental mechanism in which SCALMS maintain their stability in a coke forming environment and to extend this understanding towards designing a suitable catalyst for ethane dehydrogenation. Bismuth is used as solvent metal for the catalytically active metal, platinum, and the resulting SCALMS ($PtBi_x/Al_2O_3$ and $PtBi_x$ /SiC) were evaluated against Pt/Al_2O_3 as an equivalent conventional heterogeneous catalyst at identical reaction conditions (600, GHSV: 397 hr^{-1}, ambient pressure). In agreement with literature, EDH over Pt/Al_2O_3 shows high initial conversion although at very high selectivity toward coke, followed by rapid deactivation. In contrast, $PtBi_x/Al_2O_3$ and $PtBi_x$ /SiC show lower, but stable, conversion and stable selectivity for >10h time-on-stream. No deactivation is observed despite the fact that the cumulative amount of carbon produced on these SCALMS is comparable to that on the conventional Pt/Al_2O_3. Our results thus confirm that alloying a small amount of platinum in bismuth results in SCALMS that are both active for ethane dehydrogenation and resilient to catalytic deactivation via coking. Catalyst synthesis via various synthesis approaches (wetness impregnation, colloidal synthesis, and galvanic displacement), catalyst characterization, as well as results for catalytic performance and deactivation studies will be presented and discussed in detail in the presentation.

591f Oxidative Ethane Dehydrogenation Under Thermal and Microwave Heating over Ga/ZSM-5, Gapt/ZSM-5, and Microwave Synthesized Ga-ZSM-5

Ashley Caiola, Brandon Robinson, Sean Brown, Xiaoyan Wang, Yuxin Wang, Jianli Hu

Ethane is the second largest component of natural gas, making up to 20 vol %. Although ethane is a valuable feedstock for the production of many value-added chemicals, it is often flared at the well-head to avoid the cost associated with transporting the fuel or chance of condensation of the natural gas in the pipeline. The flaring of stranded gas adds to the global issue associated with the increased release of greenhouse emissions such as CO_2. Developing uses for this ethane to eliminate the greenhouse gas emissions and increase utilization of these valuable resources. The direct catalytic oxidative dehydrogenation of ethane (ODHE) is the use of an oxidative species as a co-reactant to produce ethylene from ethane. ODHE allows for lower reaction temperature resulting in the suppression of coke and favorable thermodynamic conditions. Using a strong oxidant such as O_2 involves the dehydrogenation of ethane to ethylene, the total oxidation of ethane, and potential formation of CO_x products which can lead to lower ethylene yield and aromatics by-products. Whereas using CO_2 as a mild oxidant can prevent the overoxidation of ethane and the metal catalyst and can be more advantageous by enhancing the equilibrium conversion obtained due to the removal of hydrogen via the water gas shift mechanism.

In this study, conventional thermal heating versus microwave heating were tested for the oxidative dehydrogenation of ethane with CO_2 as the oxidative co-reactant over conventionally synthesized Ga/ZSM-5 and GaPt/ZSM-5 and a direct microwave synthesized Ga-ZSM-5 catalyst. A 2.45GHz solid-state microwave reactor system with a high-temperature mono-mode cavity was used to supply microwave irradiation to the catalyst bed during the ODHE-CO_2. The catalytic activity of the Ga/ZSM-5, GaPt/ZSM-5 and the microwave synthesized Ga-ZSM-5 were tested under 30% ethane and 30% carbon dioxide in nitrogen, at a total flow of 100 mL/min. For the thermal fixed bed reactor, the reaction temperatures tested were 450°C, 550°C, and 650°C, whereas, for the microwave fixed bed reactor, the reaction temperature was set to 450°C all for 3 hours. The fresh catalysts were further analyzed by XRD, BET, ICP-OES, NH_3-TPD to elucidate the active metal species and catalysts prior to the ODHE reaction. The spent catalysts were characterized by TGA/DTA and TEM, to determine the effect of the reaction on the catalysts. In this study, microwave heating increased the catalytic conversion of ethane and selectivity to ethylene.

591g Portable Process Development for Methanol Synthesis from Stranded Methane

Hsiang-Sheng Chen, Nazim Muradov, Wendy Zhou, Paul Yelvington, Edwin Yik

In 2020, flared and vented natural gas (NG), derived from the petroleum industry, represented an emission level of 6.1 million metric tons CO_2 equivalent (CO_2e) in the US [1]. These associated gases are produced at oil producing wellheads that are often located in remote sites, with gas outputs that vary over their lifespans of several years, rendering the investments in pipeline infrastructure unattractive for transporting this stranded natural gas to end users. Instead, these gases are disposed, with flaring as the most common approach, to reduce methane emissions via combustion to form CO_2, which is 84 times less potent greenhouse gas over a 20-year horizon. The inherent efficiencies (~92%, [2]) of these flaring units, however, still lead to significant methane release. One alternative and revenue-generating pathway for these sources of stranded methane is through conversion to energy-dense, readily transportable methanol, which could be implemented economically if the volume and time scales of such processes can be matched to the size and lifespan of individual oil wells with associated gas.

M2X Energy Inc. has developed a prototype of a modular and portable system, which incorporates an air-breathing partial oxidation (POX) engine reformer and a methanol synthesis unit, capable of processing 75,000 standard cubic foot per day (scfd) NG to produce roughly 500 gallons/day methanol [3]. While employing similar process steps as in conventional methanol synthesis plants, the M2X system favors system simplicity and robustness over optimized yield, leading to technical tradeoffs that leverage the use of standardized process equipment more suitable for mass-production. In this work, we mention the design choices for these distributed systems and compare them to the process steps more commonly seen in conventional methanol plants. We emphasize how our design basis and process modeling were guided and enhanced by extensive bench-scale experimentation, attained through the close collaborations with our university partners at the University of Central Florida (Florida Solar Energy Center), and share our collective learnings on the unique considerations inherent within these small, modular processes intended for harvesting stranded methane.

References

1. EPA (**2022**) Inventory of U.S. Greenhouse Gas Emissions and Sinks: 1990-2020. U.S. Environmental Protection Agency, EPA 430-R-22-003.
2. Schulz, R. and de Oliviera Bredariol, T., "Flaring Emissions," *IEA, Paris,* **2022**
3. Yelvington, P. E.; Browne, J. B.; Yik, E.; Merical, K. I.; Dean, A. J.; Randolph, A. On Piston Engines as Hydrocarbon Gas Reformers for Modular, Distributed Chemical Production, *JAECS,* 13, **2023,** 100117

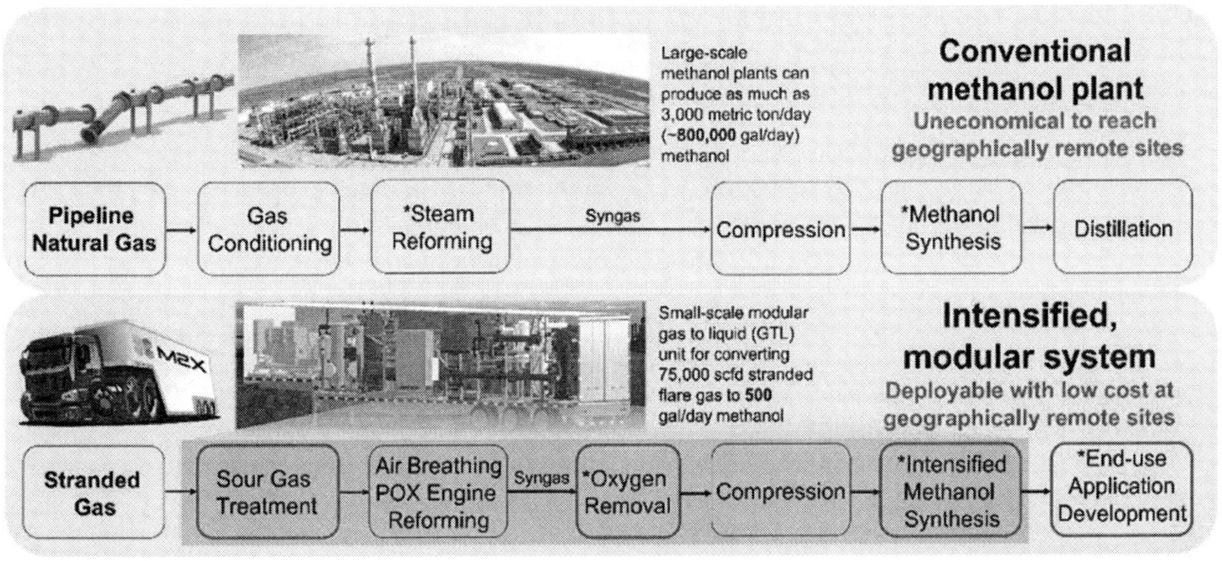
Large-scale methanol plants can produce as much as 3,000 metric ton/day (~800,000 gal/day) methanol
Conventional methanol plant
Uneconomical to reach geographically remote sites
Pipeline Natural Gas
Gas Conditioning
*Steam Reforming
Syngas
Compression
*Methanol Synthesis
Distillation
Small-scale modular gas to liquid (GTL) unit for converting 75,000 scfd stranded flare gas to 500 gal/day methanol
Intensified, modular system
Deployable with low cost at geographically remote sites
Stranded Gas
Sour Gas Treatment
Air Breathing POX Engine Reforming
Syngas
*Oxygen Removal
Compression
*Intensified Methanol Synthesis
*End-use Application Development

591h Development of a Zero-Carbon Methane Dehydroaromatization Process Design

Raid J. Hassiba, Mamoun Al-Rawashdeh, Patrick Linke

There is a growing interest in converting natural gas to products with low or zero CO_2 emissions. Methane dehydroaromatization (MDA) has received significant attention in the field of catalysis as a promising route to convert natural gas into benzene during the past two decades. To date only a few process design and integration studies have been reported for this reaction with limited analysis of catalyst coking and regeneration and with incomplete techno-economic and CO_2 emissions analyses.

This work first establishes an integrated base case MDA process design, which embeds design options proposed in the literature, such as using part of the feedstock as fuel to drive the process, accounts for catalyst regeneration and all process CO_2 emissions. The process is then progressively transformed towards a zero-carbon process by exploring alternative hydrogen separations, exploring alternative catalyst regeneration options, energy integration and switching to hydrogen enriched fuels drawn downstream of the reactor. The alternative process designs explored and compared based on the economic performance criterion Internal Rate of Return (IRR) and based on specific process CO_2 emissions per methane feed. The developed designs drive down process CO_2 emissions at each step of design exploration with the final design achieving zero-carbon emissions. Each design is simulated using commercial process simulation software and economically analyzed.

The different designs will be explained together with their economic performance and overall CO2 emissions. The results indicate profitability for all developed designs with an IRR of 14% achieved by the zero-carbon process. Besides the zero-carbon process, a second very attractive low-carbon design allows to reduce the base case process emissions by 95% while maintaining the IRR at the 18% level observed for the base case. The presented work illustrates how integrated process design and integration can lead to significant emissions reduction with little or no economic performance deterioration.

200d Renewable Blendstocks from Hazardous Industrial Wastes Via Hydrothermal Liquefaction

Harisankar S, R. Vinu

The world generates at least 2 billion tonnes of solid wastes per year, of which at least 25% are hazardous wastes from factories, industries and manufacturing plants. In a developing country like India, profit-seeking environment such as manufacturing plants, factories and small-scale industries pay little attention to sustainable practices. In most such cases, the factories produce both solid wastes as well as wastewater, which can be converted into valuable products. Such a waste-to-value conversion presents a huge business opportunity, as well as helps promote sustainable practices. It is estimated that the adoption of such sustainable practices to promote circular economy by the year 2050 will result in yearly benefits of $624 billion and a reduction in greenhouse emissions by 44%. So, seeking alternate sources to meet energy requirements is vital for a sustainable future. Hydrothermal liquefaction (HTL) is a promising thermochemical conversion process, which can convert solid wastes as well as wastewaters simultaneously to produce biofuels and chemicals. HTL uses water at near-critical condition to break down organic macromolecules present in the waste biomass to produce bio-crude, hydrochar, aqueous and gaseous phases. The bio-crude from HTL is an oxygenated analogue to petroleum crude, while hydrochar is similar to high ash-content coals.

In this work, a mixture of hazardous wastes from a tube manufacturing plant in Chennai, India, which mainly consisted of grease-rich textile, soiled gloves, antirust chemicals, and lubricant oils, were co-liquefied in a 2L batch autoclave reactor, using the wastewater produced from the same manufacturing plant to produce diesel-grade renewable blendstocks. The mixture of wastes was classified into two major fractions, viz., textile wastes and chemical wastes, and hydrothermal co-liquefaction was carried out at 300 – 350 °C, for 30 min, by varying the feedstock composition (Textile: Chemical wastes – 100:0, 75:25, 50:50, 25:75, 0:100 w/w).

Results demonstrate that the yield and quality of the bio-crude were enhanced with the addition of chemical wastes. Maximum synergy in bio-crude yield was obtained for experiment with 75:25 w/w textile:chemical wastes, at 325 °C. The HHV of the bio-crude and bio-char from this condition reached 32.4 MJ kg^{-1} and 15.3 MJ kg^{-1}, respectively, leading to an energy recovery of 92%. Ion chromatography of the aqueous phase revealed the presence of ions such as NH_4^+, Na^+, Cl^- and SO_4^{2-}. The concentration of total ions in the aqueous phase drastically got reduced with the addition of chemical wastes, although the variation in operating temperature had little effect on the ion distribution in the aqueous phase. GC/MS results suggest that the major compounds in the bio-crude are long straight-chain hydrocarbons (~35%), long-chain carboxylic acids (~30%) and long-chain amides (~20%). In other words, upgrading this bio-crude to remove the O and N heteroatoms will lead to a renewable biofuel that is very close to that of petroleum crude in properties.

The bio-crude produced from 75:25 (textile:chemical wastes) at 325 °C was further upgraded, via desalting and hydrogenation using H_2 and Ni/Al_2O_3 catalyst. The upgraded bio-crude was then distilled in the range of 150 – 380 °C to obtain HTL-diesel. The HTL-diesel was then blended with commercial diesel to produce renewable blend stocks. The blendstocks are then tested for their compatibility, efficiency and emission in an IC engine. Techno-economic analysis Considering the worldwide diesel consumption per day is nearly 4.5 billion liters, blending 10 vol.% of HTL-diesel produced from waste feedstocks with commercial diesel will save nearly $700 million worth of diesel every day.

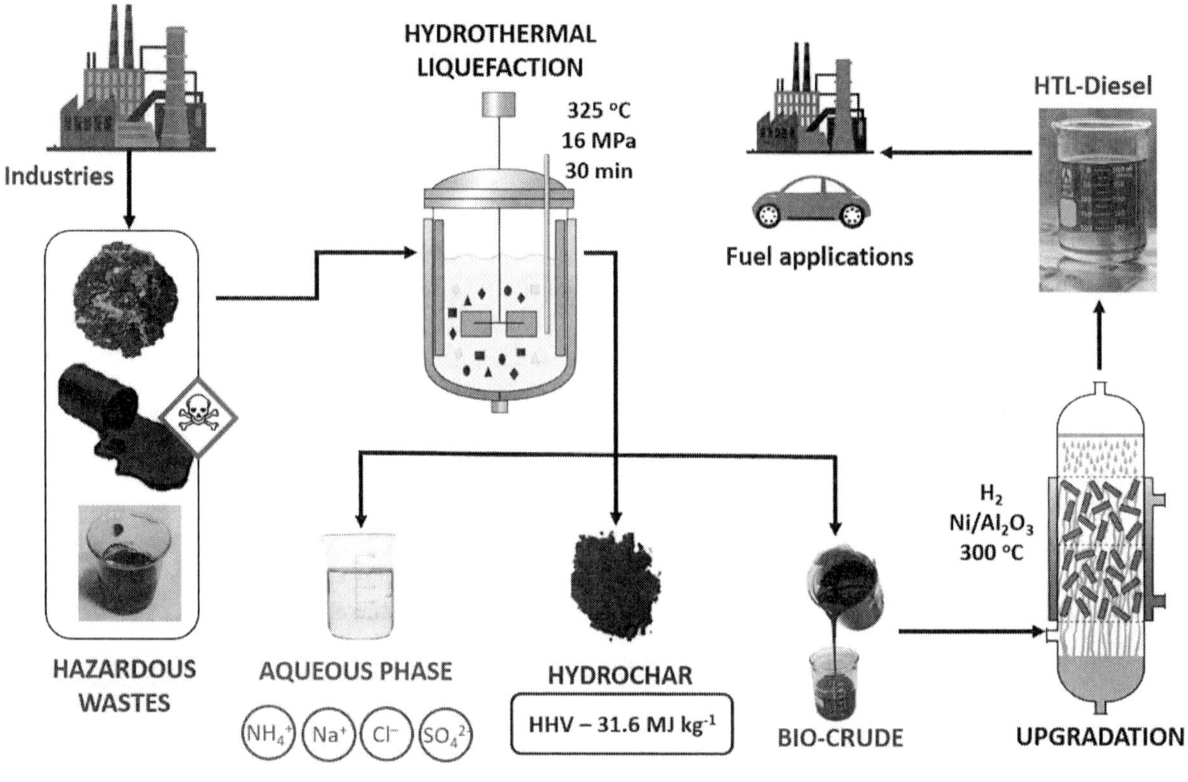
Industries
HYDROTHERMAL LIQUEFACTION
325 °C
16 MPa
30 min
Fuel applications
HTL-Diesel
H2
Ni/Al2O3
300 °C
HAZARDOUS WASTES
AQUEOUS PHASE
NH4+
Na+
Cl−
SO42-
HYDROCHAR
HHV – 31.6 MJ kg-1
BIO-CRUDE
UPGRADATION

200b Mechanical Preprocessing of Non-Recyclable Municipal Solid Wastes Towards Advanced Biorefinery Feedstock

Nepu Saha, Jordan Klinger, Wencheng Jin, Neal Yancey, Md Tahmid Islam, Toufiq Reza, Abdallah Ikbarieh, Sheng Dai, Megan Herrera

Because of accelerated urbanization and development of global economy, a large quantity of municipal solid waste (MSW) has been disposed of by the municipalities. Currently, about 50% of the total MSW generated in the United States has been utilized through various recycling, combustion, and composting technologies while the remaining 50%, often known as non-recyclable MSW (nMSW), is sent to landfills. As nMSW is physically and chemically heterogeneous, it consolidates during feeding to a biorefinery. Thus, preprocessing is required to ensure a smooth feeding process. This study focused on how mechanical pre-processing advanced the physical and chemical properties (e.g., bulk density, permeability, higher heating value (HHV), etc.) of nMSW. In addition, combustion properties were examined in terms of ignition temperature, peak heat release rate, and combustion efficiency. Results show that the variability of physical and chemical properties of nMSW can be reduced by mechanical preprocessing. For example, the variability of the bulk density of the as-received nMSW was about 17.3%, while it reduced to 5.8% when the sample was size-reduced to 2 mm. Similarly, the variability of ash and HHV reduced from 49.2% to 11.0% and from 13.4% to 4.2%, respectively. Combustion thermograms showed that the size reduction positively improves the combustion properties, while the consolidation behavior of the nMSW showed a well-fitted log-linear relationship to the permeability with applied stress. Overall, mechanical preprocessing reduced the variability of physical and chemical properties in addition to the improvement of combustion behavior of the nMSW which is one step forward toward biorefinery feedstock.

200c Process Simulation and Yield Optimization of Bio-Oil from Hydrothermal Liquefaction of Macroalgae(Kelp) Using Mixtures of Model Compounds.

Micheal Asama, Fernando Resende, Aaditya Khanal

ABSTRACT

Hydrothermal liquefaction (HTL) of macroalgae (kelp) has been recently receiving significant attention as an environmentally friendly process for producing renewable fuels. However, Due to the intricacy of the reaction mechanism and variety in the organically rich wet feedstock being evaluated for HTL, it poses challenges to the accurate prediction of product yield. Estimating the quantity of the products is vital to effectively develop and optimize industrial-scale HTL operations. To address this challenge, a batch simulation model was developed to predict the four phases of product formation: aqueous, gas, solid, and oil, using a macro-algae feedstock modeled as mixtures of carbohydrates, protein, and lipids. The model considered the effects of time, temperature, pressure, and feedstock loading on the product yields and was validated using experimental data obtained from the hydrothermal liquefaction of macroalgae. This study used response surface methodology (RSM) to evaluate and optimize bio-crude production using experimental data collected from the literature and rigorous process simulations. We developed a mathematical model using data conducted at numerous temperatures, pressure, residence times, and feedstock loading to predict the statistical correlation between actual predictors and the response variables. A central composite design was employed to generate experimental runs, which are then simulated and statistically analyzed. The Analysis of Variance (ANOVA) of the means showed that the selection of operational parameters could significantly affect the bio-oil yield. The response surface model obtained was statistically significant ($p<0.05$), indicating that the model explains a large proportion of the variability in the bio-oil yield. The optimal process conditions for maximum bio-oil yield were also identified. The response surface model was validated against the simulated data obtained from the simulator with satisfactory accuracy. This study successfully optimizes the HTL of macroalgae (kelp) for maximum bio-crude yield, using RSM and ANOVA analysis. The results highlight the significance of identifying the key process parameters that significantly affect the response variable, which is essential for optimizing the process. The methodology used in this study can be applied to similar processes in the future.

200a Renewable Liquefied Natural Gas As Marine Fuel: Process Simulation-Based Analysis of Different Production Routes Including Energetic, Environmental and Economic Aspects

Elena Barbera, Andrea Mio, Michele Adinolfi, Maurizio Fermeglia, Alberto Bertucco

Maritime transportation contributes to about 3% of global anthropogenic greenhouse gas (GHG) emissions, adding up to 1.5 billion tons of $CO_{2,eq}$ per year, including emissions from fuels production as well as those derived from their combustion. These emissions need to be reduced by at least 50% in absolute value by 2050, to contribute to the ambitions of the Paris Agreement. Compared to conventional fuels such as Marine Gas Oil (MGO) and Heavy Fuel Oil (HFO), using Liquefied Natural Gas (LNG) can lower GHG emissions by 20-25%, thanks to the low sulfur and nitrogen content, and to the favorable hydrogen-to-carbon ratio that reduces particle emissions. For this reason, a transition towards LNG engines has been occurring in the latest years. However, to effectively reduce GHG emissions, zero-carbon fuels produced from renewable sources are to be envisaged.

LNG can in fact be produced from renewable sources according to different strategies. Electro-LNG, or *e-LNG*, can be obtained from renewable electricity, which is exploited in the production of green hydrogen by means of water electrolysis. Hydrogen can then be reacted with carbon dioxide captured from the air or from some other emission sources, allowing the production of synthetic methane. In the case of capture from the air, the CO_2 balance is neutral, because the CO_2 produced by the combustion of e-LNG is sequestered and reused as a process reactant, thus closing the cycle.

Alternately, LNG can be produced by anaerobic digestion of organic matter, such as urban, agricultural and industrial organic waste (*bio-LNG*). In this case, the biogas obtained from anaerobic digestion (mainly a mixture of CH_4 and CO_2), has to be upgraded to meet the appropriate methane purity. The resulting CO_2 balance is neutral, as the CO_2 obtained as a secondary product of the anaerobic digestion process and produced by the combustion of bio-LNG comes from the photosynthesis process.

This work aims at analyzing different renewable LNG synthesis pathways at industrial scale, and at comparing them in terms of their energetic efficiency, economic profitability, and environmental impacts by means of acknowledged quantitative performance indicators which are respectively the Energy Return on Energy Invested (EROEI), the Levelized Cost of Energy (LCOE), and the Life Cycle Assessment (LCA). The estimated bunkering capacity of an average-size port in North-East Italy will be taken as reference. In order to have a fair and robust comparison among the different processes investigated, the conceptual process design and the material and energy balances required to evaluate the aforementioned indicators have been developed by means of process simulation.

Results indicate that, due to the large energetic burden of green hydrogen production by means of water electrolysis, the most convenient pathway appears to be bio-LNG production, whose levelized cost of production is however 2-3 times larger than that of the fossil counterpart. In this case, however, the main issue is related to feedstock availability to meet the required production capacity.

200f Catalytic Fast Hydropyrolysis of Beetle Killed Trees- Pine *(pinus taeda), Ash(fraxinus) and Redbay (persea borbonia)* for the Production of Transportation Biofuel

Oluwanisola Makinde, Fernando Resende

Tree-killing insects are responsible for large-scale deforestation across North America. For instance, Emerald ash borers are responsible for infesting and killing several ash trees especially in Louisiana and the East Texas region and this beetle infested trees becomes a threat to public safety when they fall off. However, these trees can be converted into useful products such as biofuels.

This study investigates the potential of using beetle-killed trees as feedstocks for biofuel production via thermochemical processes called pyrolysis and hydropyrolysis. We report the effects of varying catalysts and reaction conditions on product yield. We observed that non-catalytic fast pyrolysis produced mainly oxygenated compounds, however, upgrading to catalytic fast pyrolysis (CFP) using HZSM-5 promotes deoxygenation reactions, turning oxygenated volatiles into hydrocarbon fuels, mainly aromatics such as benzene, toluene, and xylene (abbreviated BTX).

We evaluated the influence of catalyst/biomass ratio of 20:1 which is an optimum ratio at temperature of 600°C, on the pyrolysis product yield and compared the results obtained for the different feedstocks. Our results show a good yield of benzene, toluene, and xylene for the different biomass under study, with pine producing the highest hydrocarbon yield of 76% based on percentage area.

We will report results for catalytic hydropyrolysis (CHP) of these beetle-killed trees using HZSM-5, NiMo/HZSM-5 and NiRe/HZSM-5, targeting high yields of hydrocarbons such as BTX (benzene, toluene, and xylene), alkanes, and cycloalkanes, which are some of the major components of jet-fuels and are important raw materials for the petrochemical industry. The range of temperature of this study is 400-600^{0}C,and the range of pressure is 20-40 bar.

Compared to catalytic fast pyrolysis(CFP), the hydrogen environment in catalytic hydropyrolysis (CHP) can potentially reducing coking, increase yields of aromatics and shift the product composition towards alkanes and cycloalkanes. It is expected that HZSM-5 catalyst should produce more BTX compounds, while the NiMo/HZSM-5 and NiRe/HZSM-5 catalyst produce alkanes and cycloalkanes. The optimum pressure and temperature for high product yield will be determined.

The utilization of beetle-killed trees as a feedstock for biofuel production also has environmental benefits, such as reducing waste wood in forests and mitigating greenhouse gas emissions. Overall, this study provides valuable insights into the use of beetle-killed trees for biofuel production and highlights the importance of exploring alternative feedstocks for sustainable transportation energy.

200g Engineering the Kinetics of Acid Hydrolysis of Corn COB to Glucose for BIO-Ethanol Production

Okechukwu Onyelucheya, Jude C Obijiaku, Chioma M Onyelucheya

Corn cob as a renewable, cheap and abundant lignocellulosic material has been found to be a potential source of fermentable sugars, which can serve as a substrate for value added products such as ethanol. Our research project studied the glucose production from corn cob using phosphoric and nitric acids hydrolysis at 130^0C, at an acid concentration range of 1-5wt% and at time intervals between 18 to 90mins. The high hemi-cellulosic content of the substrate was confirmed through a proximate constituent's characterization. Seaman's model and the Two-fraction model were useful in studying the kinetics of the glucose production in the hydrolysates and having the Two-fraction model give a better fit with R^2 approximately equal to 1 indicated suitability for use in engineering the kinetics of the acid hydrolysis process. Within 42 mins, glucose concentration as much as 0.038mg/ml was produced when 5wt% H_3PO_4 was used as catalyst for the hydrolysis, with 0.4% susceptibility to hydrolysis. And within 77mins, up to 0.054mg/ml glucose concentration was produced for the nitric acid hydrolysis with 2wt% HNO_3, with 0.6% susceptibility to hydrolysis. Comparatively, nitric acid proved to be a more efficient catalyst for the acid hydrolysis of the corn cobs and provided great insight to the engineering of the kinetics of the acid hydrolysis process for greater glucose yield and in turn higher potential for bio-ethanol production capacity.

417a Determination of the Cathode Degradation Mechanism in Hydroxide Exchange Membrane Fuel Cells

Catherine M. Weiss, Brian P. Setzler, Yushan Yan

As the climate crisis worsens, decarbonizing the global economy has become a matter of increasing urgency. Today, the transportation sector accounts for roughly a quarter of all global greenhouse gas emissions. Promisingly, Proton Exchange Membrane Fuel Cells (PEMFCs) are emerging as a commercially viable power source for low-carbon intensity transportation applications. However, PEMFC costs today far exceed the DOE target, limiting their adoption. Hydroxide Exchange Membrane Fuel Cells (HEMFCs) hold the promise to be a cost-effective alternative to PEMFCs due to the potential for lower stack capital costs. Specifically, in alkaline, several inexpensive non-Platinum Group Metals (PGM) are stable and active catalysts, and stack bipolar plates can be manufactured using less expensive materials. Moreover, the Hydroxide Exchange Membranes (HEM) display lower gas crossover compared to PEMs. Currently, low chemical and mechanical durability is a key barrier to the commercialization of HEMFCs. However, the mechanisms of degradation in HEMFCs have not been thoroughly investigated and are not well understood thus far.

We examine the degradation of the cathode electrode in HEMFCs due to three possible contributions: alkaline instability of the ionomer, carbon corrosion of the catalyst support, and cleavage of the ionomer backbone due to oxygen radicals. A HEMFC constant current held at 0.5 A cm^{-2} under pure H_2 and O_2 had a nonlinear voltage drop of 0.37 V over 180 h. The results from the fuel cell durability experiment were compared to data from durability tests designed to probe each mechanism. Using identical membrane electrode assemblies, we performed a HEM H_2 Pump Hold to evaluate ionomer alkaline stability, a High Voltage Hold to evaluate carbon corrosion of the support, and an Oxygen Limited Hold to evaluate radical oxidation of the ionomer. In the alkaline environment, the carbon support is active for 2 e- ORR, producing peroxide which is converted into hydroxyl and peroxyl radicals. At a constant current density of 0.5 A cm^{-2}, we show that degradation is primarily driven by radical degradation of the ionomer, accelerated by the carbon support. Radical-induced ionomer chemical degradation decreases the ionomer mechanical strength, leading to catalyst layer delamination. By eliminating the carbon support and adding ruthenium, a potential radical scavenger, the HEMFCs achieved a constant current hold at 0.5 A cm^{-2} for 190 h with a degradation rate of 9 $\mu V\ h^{-1}$.

417b Dynamic Modeling and Control of a PEM Electrolyzer for Solar Photovoltaic Power Smoothing

Yifan Wang, James Fenton, Paul Brooker

The variability and intermittency of solar photovoltaic (PV) electricity under increasing renewable penetration can lead to frequent and steep ramping operations of conventional fossil fuel power plants. Hydrogen production via water electrolysis can serve as a controllable load and provide both short and long duration energy storage to reduce grid fluctuations and improve the resiliency of the grid system. The high efficiency and quick dynamic response of a polymer electrolyte membrane (PEM) electrolyzer can be used to both generate hydrogen and serve as a controllable load [1,2]. The capacity factors of both PV field and PEM electrolyzer can affect the performance of PV smoothing on the cloudy day. The generated green hydrogen must be stored under high pressure for subsequent utilizations (e.g., hydrogen fuel cell electric vehicle refueling). PEM electrolyzers directly producing high pressure hydrogen are more energy efficient than ambient electrolyzers followed by mechanical compression, but higher pressure leads to more undesired hydrogen permeation back to the oxygen side, which can cause lower hydrogen production efficiency, cell degradation, and safety risk of hydrogen explosion limit [3,4]. Electrolyzer operational pressure optimization under fluctuating electricity (i.e. current density) is developed for an integrated hydrogen system (renewable solar with green hydrogen production and storage). Different PV smoothing scenarios and varied pressure operation conditions of the PEM electrolyzer are investigated.

A high-fidelity dynamic model of a PEM electrolyzer cell/stack is developed with consideration of one-dimensional ('through-plane') mass/heat transfer coupled with electrochemical kinetics. The electrochemical reactions are modeled using Butler-Volmer kinetics with varying surface molar concentrations of components. Maxwell-Stefan diffusion equation is used to calculate the gas-phase species molar concentration in the backing and catalyst layers. Hydrogen permeation through membrane by diffusion and convective transport as a function of operational pressure is assessed. To evaluate the transient response of the PEM electrolyzer stack, real-time PV data based on a solar farm from Orlando Utilities Commission (OUC) is smoothed by the electrolyzer using a power signal control algorithm. The developed PV smoothing control algorithm and the electrochemical dynamic stack model show the effectiveness of an electrolyzer in smoothing the PV signal to increase the grid stability and flexibility.

[1] Kumar, S.S. and Himabindu, V., 2019. Hydrogen production by PEM water electrolysis–A review. Materials Science for Energy Technologies, 2(3), pp.442-454.

[2] Götz, M., Lefebvre, J., Mörs, F., Koch, A.M., Graf, F., Bajohr, S., Reimert, R. and Kolb, T., 2016. Renewable Power-to-Gas: A technological and economic review. *Renewable energy, 85*, pp.1371-1390.

[3] Trinke, P., Bensmann, B., Reichstein, S., Hanke-Rauschenbach, R. and Sundmacher, K., 2016. Hydrogen permeation in PEM electrolyzer cells operated at asymmetric pressure conditions. Journal of The Electrochemical Society, 163(11), p.F3164.

[4] Scheepers, F., Stähler, M., Stähler, A., Rauls, E., Müller, M., Carmo, M. and Lehnert, W., 2020. Improving the efficiency of PEM electrolyzers through membrane-specific pressure optimization. Energies, 13(3), p.612.

417c Phosphonic Acid Ionomers for Use As Free Radical Scavengers in Proton Exchange Membrane Fuel Cells

Alisa Chakraborty, Christopher Arges

Proton-exchange membrane fuel cells (PEMFCs) are an electrochemical energy conversion device that has no carbon emissions provided it uses green hydrogen as a fuel. During fuel cell operation, the PEM undergoes mechanical, thermal, and chemical degradation. The chemical degradation is attributed to reactive oxygen species (ROS) that are generated in situ through both chemical and electrochemical pathways during fuel cell operation. Hydrogen peroxide (H_2O_2) is formed as an intermediate and reacts with trace transition metal ions like Fe^{2+} or Fe^{3+} to generate ROS. The ROS mainly comprise of hydroxyl and hydroperoxyl radicals and these radicals initiate oxidative degradation of both the PEM backbone and the side chains that contain ionic groups that are essential for proton conduction. In this study, Nafion-phosphonic acid ionomer blends were prepared into composite PEMs. The phosphonic acid ionomers were polyvinyl phosphonic acid (PVPA) and polytetrafluorostyrene phosphonic acid (PTFSPA). Using fluorescence spectroscopy, we observe scavenging of ROS species through the incorporation of phosphonic acid ionomers. The scavenging ability was observed to be on par, or to exceed, cerium oxide nanoparticles. The distribution and mobility of the phosphonic acid ionomer within Nafion were assessed using STEM EDX mapping. The incorporation of phosphonic acid ionomers, in some cases, were shown to improve membrane proton conductivity.

417d Enhanced Power Management for Direct Methanol Fuel Cell/Battery Hybrid System By Improved Cascade Control Structure

Zuhair Alyousef, Majed Aljeshi

BACKGROUND

Effective control structure for fuel cell hybrid systems is one of the critical milestones in the design phase that needs due attention. In the applications where the load connected to the cell is expected to change its demand in a frequent basis, the device might not be capable in some occasions to satisfy the load requirement. This is attributed to the fact that fuel cells' electrical outputs are often fluctuating, and their dynamics are relatively slow during some transient periods such as start-up and shutdown.

In view of the fuel cell job of powering the connected loads, its integration in a hybrid system is essential to overcome the aforementioned shortcoming in the applications with changing loads. The literature has a good amount of work treating the hybrid systems consisting of fuel cells and batteries, supercapacitors or other storage devices. There are two main configurations available for hybrid systems (i) the series structure where the battery is the main source of energy to the load, and (ii) the parallel form where both devices feed the connected load. The first structure is considered less challenging since the cell feeds the battery which in turn provides the power to the load. On the other hand, the second configuration requires more considerations in the design and implementation of control strategy. To this end, the literature needs more research works investigating optimal designs of parallel systems to make them more efficient. In this work we provide a proposed parallel configuration structure for direct methanol fuel cell (DMFC)/Li-ion battery hybrid system that maximizes the dependance on DMFC and reduces the use of battery. We address the benefits and traits of the proposed architecture along with selected results.

PARALLEL HYBRID SYTEM CONFIGURATION CONCEPT

The parallel configuration utilizes a power management system (PMS) in a form of a supervisory control strategy. In this system, the electrical signal of the load is sent to PMS and the controller makes the decision whether to deliver the power by DMFC solely or by the combination of fuel cell and battery.

The signature of this work lies in the utilization of an improved cascade control structure that we call a "dual threshold algorithm" installed to regulate the liquid-level of methanol solution tank feeding the oxidation reaction species in DMFC. As discussed briefly in the next part, this control logic extends its service beyond the level management and leads to a stable temperature output of the cell that can be manipulated -intermittently- based on the user specification. The ultimate advantage is the ability to change the power of the DMFC from one state to another as a result of the temperature availability as a manipulated variable. In this fashion, the fuel cell can shift its power to satisfy the changes in load demands. However, this can happen in some periods during the course of operation where PMS will only use the feed of DMFC and this leaves the battery on sleep mode.

IMPORTANCE AND ADVANTAGES

The parallel hybrid configuration requires more investigation and research work to further advance the performance of such system. We argue that with a proper power management system organizing the electrical delivery to the connected load, several advantages could be realized. First, the dependance on the fuel cell could be maximized such that the battery life can be extended. Hence, the expenses associated with battery replacement become lower. Second, the battery can be used as a backup to compensate for the power in the event the fuel cell fails in satisfying load requirement. Thus, the system size is reduced since the battery pack needed would be less. Another aspect of pivotal role is the improvement of system efficiency where there is a good flexibility of optimizing the operational condition for both the fuel cell and battery pack.

CHALLENGES OF IMPLEMENTATION

Figure 1 demonstrates the cascade control structure used to regulate the level in the methanol solution tank of DMFC. The symbols LT, TT respectively designate the level transmitter and temperature transmitter. The level controller (master) is denoted by LC while TC signifies temperature controller (slave). The level set point is denoted by L_{sp}. The fan is the final control element operated by TC. The output of the master proportional-integral (PI) controller is given by

$$T^{AFC} = T_{bias} + K_p\, e(t) + K_i \int e(\theta)d\theta \quad (1)$$

where T^{AFC} is the temperature of the anode flow channel, T_{bias} is the bias temperature, K_p and K_i are respectively the proportional and integral gains, while $e(t) = L_{sp} - L(t)$ is the feedback error. It is evident from this law that the operational temperature of the cell (approximated to be T^{AFC}) is always prescribed by the master LC. This controller in turn has a single objective of tracking the set point L_{sp}. Thus, the temperature is continuously manipulated for the sole purpose of level management. The ultimate result is non-ability to operate the cell at the desired output power which is directly affected by temperature.

In view of cascade control mechanism, if the fuel cell is integrated in a hybrid system, the output power of the cell cannot be altered or shifted from one state to another since the temperature is not free for change. Consequently, upon load changes the battery has to be invoked in most occasions which violates the objective we try to achieve of maximizing the dependance on DMFC.

MODIFICATION AND IMPROVEMENT

A modified cascade control architecture detailed in [1] is proposed to improve the electrical performance of the cell. Its main idea is to make the tank operating within an upper and lower threshold enclosing a medium set point. The law of the modified cascade control is given by

$$T^{AFC} = T_{bias} + f(t)\, [\, K_p\, e(t) + K_i \int e(\theta)d\theta\,] \quad (2)$$

where $f(t)$ is a switching function adopting the binary value of 0 or 1. This modified version works in a fashion that when the level is within the threshold band, the switching function becomes 0 and the output of the controller is reduced to the bias temperature that can be freely selected by the user for any objective, such as changing the output power. On the other hand, if either one of the thresholds is violated, $f(t)$ becomes 1 and the controller is returned to standard cascade so that the level is returned to the medium set point. Our work presented in [1] demonstrates via simulation environment that the cell can operate in periods of significant time where the master and slave controllers are decoupled ($f(t)=0$) and the power can be shifted successfully from one state to another.

This improvement shows that if the cell is integrated in a hybrid system, the battery usage becomes less frequent and extends its lifetime.

SELECTED RESULTS

This work is based on the DMFC/Li-ion battery hybrid system where we introduce the power management system in a simulation environment using MATLAB and Simulink software suite. We utilize a detailed dynamic modeling of open-cathode DMFC system that is developed by our research group while a widely used Li-ion battery model is imported from Simulink.

The reader should observe that there are several case scenarios for this hybrid system mode, where in some occasions the master and slave controllers are decoupled ($f(t)=0$), while in other events they are coupled ($f(t)=1$). Here and for the purpose of brevity, we show one example of the results, and we leave other cases for more detailed manuscripts.

Figure 2 shows the simulation results in the event where the master and slave controllers are decoupled with a switching function of 0.

The upper plot demonstrates the signal of the power requested by the load, the second is the output power from the fuel cell, the third is the battery power (both 2 and 3 as dictated by PMS), while the fourth is the total summation of the fuel cell and battery.

As seen from the figure, the requested power by the load is initially at 47 W and that is totally delivered by the fuel cell. A step change in the load from 47 W to 50 W takes place at 2000 seconds and that is also delivered by the cell while the battery is on sleep mode. The success of the fuel cell to change its output is attributed to the ability to change temperature from one state to another as given by Equation (2) when $f(t)=0$.

REFERNCES

[1] Zuhair A Al-Yousef, Shyam P Mudiraj, and Oscar D Crisalle. Improved cascade control

structure for water and thermal management in open-cathode direct methanol fuel cells. In

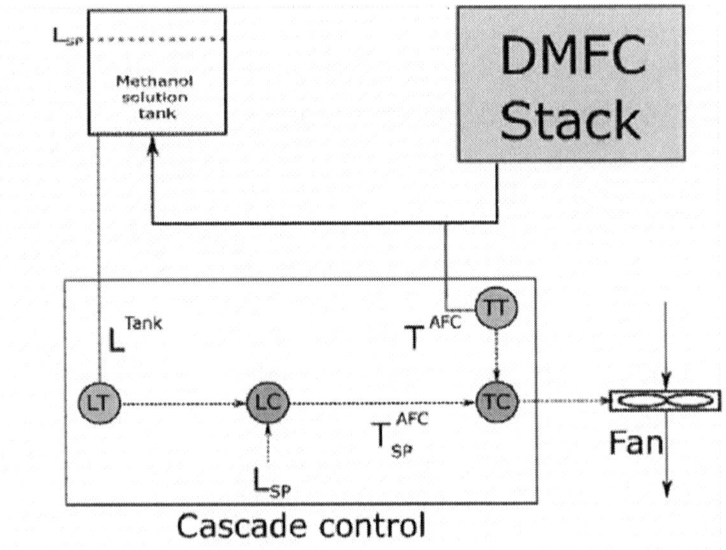

Figure 1: Schematic diagram of cascade control structure for liquid-level management of DMFC.

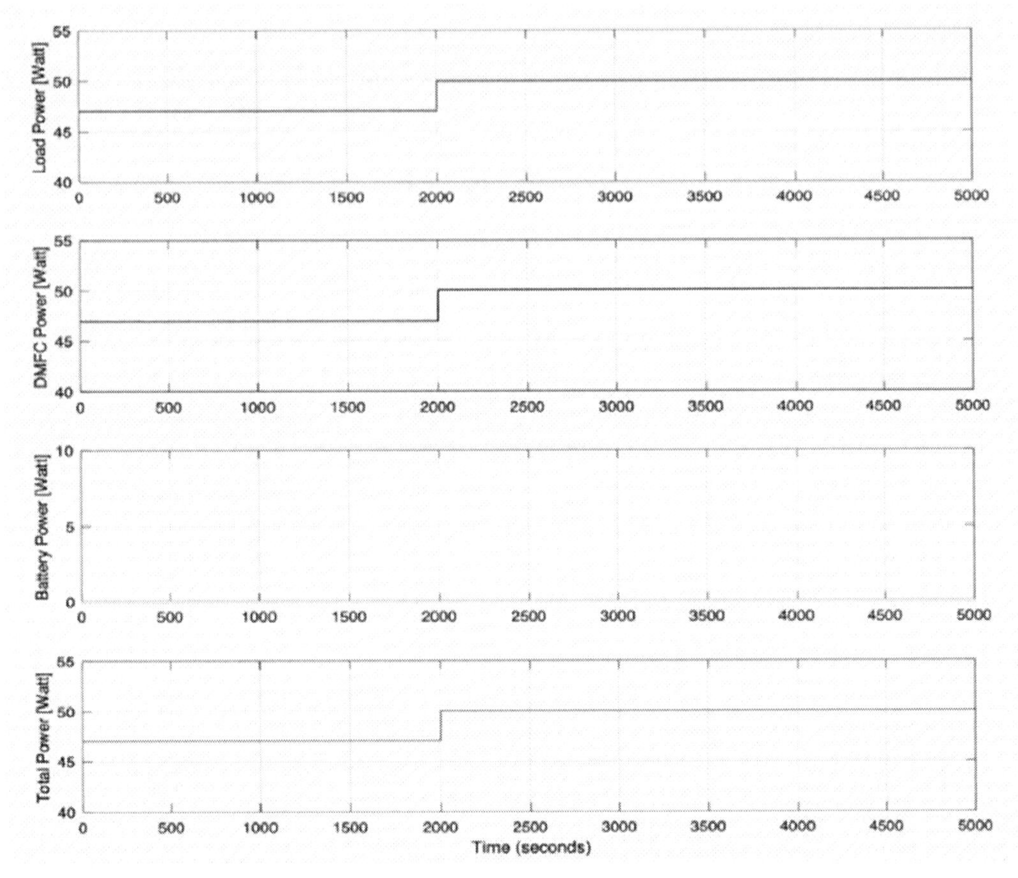

Figure 2: Simulation results of PMS outputs to shift the load power from 47 to 50 W.

417e Analysis of Mixed Potential at the Cathode of a Direct Methanol-Hydrogen Peroxide Fuel Cell for Enhanced Performance

Kritika Sharma, Vijay Ramani

Direct methanol fuel cells (DMFCs) are promising energy conversion devices that can generate electricity from the electrochemical oxidation of methanol. However, the use of traditional oxidants, such as oxygen, can lead to various challenges, such as catalyst poisoning and incomplete fuel oxidation. Hydrogen peroxide (H_2O_2) has emerged as an attractive alternative oxidant for DMFCs due to its high energy density and non-toxic nature. The use of H_2O_2 as an oxidant in fuel cells offers several advantages over traditional oxygen-based systems, including improved safety, reduced corrosion, increased efficiency and use in O_2 free environment.

Direct methanol hydrogen peroxide fuel cells (DMHPFC) are an important alternative to hydrogen-fed polymer electrode membrane fuel cell (PEMFCs) and anion exchange membrane fuel cells (AEMFCs) due to methanol's high energy density compared to that of hydrogen and superior kinetics of hydrogen peroxide reduction reaction as compared to Oxygen. A unit volume of Hydrogen gas stored at a pressure of approximately 69 MPa corresponds to volume-specific energy density of roughly 2.1 kWh/l while 39% volume of aqueous methanol and 41% volume of aqueous hydrogen peroxide corresponds to volume-specific energy density of 9.2 kWh/l. So, a DMHPFC's energy density is approximately four times higher than hydrogen fuel cell's energy density and is comparable to gasoline which contains roughly 9.2 kWh/l of available bond energy. We have achieved sustained operation of the DMHPFC with a peak power density of 0.8 W cm^{-2} and a power density of > 0.5 W cm^{-2} at 1.0 V. However, the system exhibits a relatively low Faradaic efficiency of 50% due to various overpotentials.

In DMHPFCs, the cathode overpotential is critical and determines the overall performance of the fuel cell. The cathode reaction involves hydrogen peroxide reduction, which is a complex process because of the parasitic evolution of O_2 by the decomposition of H_2O_2. The O_2 evolution results in a mixed potential at the cathode due to the occurrence of both the Hydrogen peroxide reduction reaction, (HPRR) (E_0 = 1.77 V versus standard hydrogen electrode [SHE]) and the oxygen reduction reaction (ORR) (E_0 = 1.23 V versus SHE), lowering the overall cell potential. The O_2 surface coverage reduces the available sites for the HPRR, effectively deactivating the catalyst. Electrocatalysts exhibiting a combination of high HPRR activity and selectivity (by inhibiting H_2O_2 decomposition, and hence inhibiting ORR would result in higher faradic efficiency and have been the subject of sustained interest. Another challenge at the cathode is the crossover of methanol from the anode to the Pt-based Hydrogen peroxide cathode. This mixed potential due to methanol crossover can have a significant impact on the performance of the cathode and the overall efficiency of the DMHPFC.

Understanding and controlling the mixed potential at the cathode in DMHPFCs is crucial for optimizing the cell performance and improving its efficiency. The mixed potential at the cathode depends on various factors such as the concentration of methanol and hydrogen peroxide, pH, temperature, membrane, and electrode materials. In this work, we will provide an overview of the mixed potential theory and its significance in the DMHPFCs. We will also discuss the factors influencing the mixed potential at the cathode and implications for the cathode reaction, and strategies for mitigating its effects.

570a Carbon Catalyzed Decarbonization of Natural Gas for Clean Hydrogen and Solid Carbon

Randall Vander Wal, Mpila Nkiawete

Hydrogen is envisioned as the energy carrier (fuel) of the future and is a crucial feedstock for various manufacturing industries. Today ~ 10 million metric tons (MMT) are produced in the U.S, ~ 95% of this is from steam reforming of methane (SMR), for use in oil refineries, methanol and ammonia production [1]. SMR produces 13.7 kg CO_2 (equiv.)/kg of net hydrogen [2] and consumes 19.8 liters of water per kg of hydrogen [3]. In contrast, thermo-catalytic decomposition (TCD) of natural gas does not produce CO_2, nor does it consume water resources [4]. The energy requirement for methane cracking process (37.8 kJ/mole of H_2) *is substantially less* than that for steam reforming (74.8 kJ/mole of CH_4) [5-7]. Finally, the carbon stripped from the natural gas is in solid form – ideal for carbon sequestration [8,9].

Decarburization of natural gas (NG) represents a path towards the hydrogen economy till renewable H_2 generation reaches scale. The solid (elemental) carbon produced not only presents a tremendous advantage in carbon capture and storage [8,10-13] but has potential commercial value [14], as additive for composites, structural applications and even for soil amendment [15,16]. Given its purity, it is useable as electrode materials in energy storage applications [1,17] or metallurgical processing [18].

The barrier to TCD is maintaining catalytic activity [1-6]. TCD requires a catalyst to overcome the high thermodynamic barrier for NG decomposition to H_2 and solid carbon [5]. Carbon itself is an excellent catalyst, cheap, and tolerant of NG impurities that can poison (deactivate) a metal catalyst [2,4,7,8]. Thus, the depositing carbon could potentially serve as catalyst [19].

In this study, a synthetic natural gas blend diluted in an argon carrier served as the initial feed. The test matrix encompassed a series of temperatures (900 – 1,100 °C) and reaction durations (1 – 12 hrs.). TCD measurements were performed using a hot wall reactor and flat substrates of silicon and quartz. After deposition for varied durations, substrates were removed for film thickness determination. TCD rates were measured by deposit thickness, using scanning electron microscopy (SEM). As kinetic rates are mapped, they are related to active site number for each of the two processes, TCD and regeneration, in order to provide a common connection

The nanostructure of the deposit was accessed using transmission electron microscopy (TEM) and Raman spectroscopy. Nanostructure was assessed to identify correlations with kinetic rates. Nanostructure is an easier metric with which to gauge carbon catalyst activity and deposition rate, with comparison to chemisorption followed by XPS, an approach more definitive than TPD. For reference there is considerable literature precedence using nanostructure for interpretation of soot oxidative reactivity [20-22]. Here nanostructure is also used to gauge the efficacy of initial carbon catalysts prior to TCD testing. At selected stages during TCD, samples were subjected to activated chemisorption in preparation of active site measurement. Active sites were quantified by X-ray photoelectron spectroscopy (XPS) with deconvolution to identify oxygen functional groups. The differences in oxygen content between nascent and partially oxidized samples quantifies the number of active sites created while the difference in oxygen content across the two carbons can be attributed to the nanostructure differences (and corresponding active site number) between the two carbon samples.

References

1. Dagle, R. A., Dagle, V., Bearden, M. D., Holladay, J. D., Krause, T. R., & Ahmed, S. (2017). *An Overview of Natural Gas Conversion Technologies for Co-Production of Hydrogen and Value-Added Solid Carbon Products* (No. PNNL-26726; ANL-17/11). Pacific Northwest National Lab. (PNNL), Richland, WA (United States); Argonne National Lab. (ANL), Argonne, IL (United States).

2. Abbas, H. F., Daud, W. W. (2010). Hydrogen production by methane decomposition: a review. *International Journal of Hydrogen Energy*, *35*(3), 1160-1190.
3. Spath, P. L., and Mann, M. K., Life cycle assessment of hydrogen production via natural gas steam reforming. (2001). Technical Report NREL/TP-570-27637.
4. Muradov N. Thermocatalytic CO2-free production of hydrogen from hydrocarbon fuels. U.S. DOE Hydrogen Program Review. U.S.: Department of Energy (DOE); 2002. NREL/CP-610-32405.
5. Amin, A. M., Croiset, E., Epling, W. (2011). Review of methane catalytic cracking for hydrogen production. *international journal of hydrogen energy*, *36*(4), 2904-2935.
6. Torres, D., Pinilla, J., & Suelves, I. (2018). Co-, Cu-and Fe-Doped Ni/Al2O3 Catalysts for the Catalytic Decomposition of Methane into Hydrogen and Carbon Nanofibers. *Catalysts*, *8*(8), 300.
7. Muradov, N. (2001). Hydrogen via methane decomposition: an application for decarbonization of fossil fuels. *International Journal of Hydrogen Energy*, *26*(11), 1165-1175.
8. Ashik, U. P. M., Daud, W. W., Abbas, H. F. (2015). Production of greenhouse gas free hydrogen by thermocatalytic decomposition of methane–a review. *Renewable and Sustainable Energy Reviews*, *44*, 221-256.
9. Gaudernack, B., Lynum, S. (1998). Hydrogen from natural gas without release of CO2 to the atmosphere. *International journal of hydrogen energy*, *23*(12), 1087-1093.
10. Muradov, N. (2001). Hydrogen via methane decomposition: an application for decarbonization of fossil fuels. *International Journal of Hydrogen Energy*, *26*(11), 1165-1175.
11. Muradov, N. (2000, May). Thermocatalytic CO2-free production of hydrogen from hydrocarbon fuels. In *Proceedings of the 2000 Hydrogen Program Review, NREL/CP-570-28890*.
12. Muradov, N., Chen, Z., & Smith, F. (2005). Fossil hydrogen with reduced CO2 emission: modeling thermocatalytic decomposition of methane in a fluidized bed of carbon particles. *International journal of hydrogen energy*, *30*(10), 1149-1158.
13. Muradov, N. Z. (1998). CO2-free production of hydrogen by catalytic pyrolysis of hydrocarbon fuel. *Energy & Fuels*, *12*(1), 41-48.
14. Parkinson, B., Tabatabaei, M., Upham, D. C., Ballinger, B., Greig, C., Smart, S., & McFarland, E. (2018). Hydrogen production using methane: Techno-economics of decarbonizing fuels and chemicals. *International Journal of Hydrogen Energy*, *43*(5), 2540-2555.
15. De Falco, M., Basile, A. (Eds.). (2015). *Enriched Methane: The First Step Towards the Hydrogen Economy*. Springer.
16. Muradov, N. Z., & Veziroğlu, T. N. (2005). From hydrocarbon to hydrogen–carbon to hydrogen economy. *International Journal of Hydrogen Energy*, *30*(3), 225-237.
17. Lane, J. M., & Spath, P. L. (2001). Technoeconomic analysis of the thermocatalytic decomposition of natural gas. *Colorado USA: Department of Energy (DOE) US.*
18. Parkinson, B., Tabatabaei, M., Upham, D. C., Ballinger, B., Greig, C., Smart, S., & McFarland, E. (2018). Hydrogen production using methane: Techno-economics of decarbonizing fuels and chemicals. *International Journal of Hydrogen Energy*, *43*(5), 2540-2555.
19. Vander Wal, R., & Nkiawete, M. (2020). Carbons as catalysts in thermo-catalytic hydrocarbon decomposition: a review. *C*, *6*(2), 23.
20. Vander Wal, R. L., & Tomasek, A. J. Soot nanostructure: dependence upon synthesis conditions. *Combustion and Flame*, 2004, *136*(1-2), 129-140.
21. Yehliu, K., Vander Wal, R. L., Armas, O., & Boehman, A. L. Impact of fuel formulation on the nanostructure and reactivity of diesel soot. *Combustion and Flame*, 2012, *159*(12), 3597-3606.
22. Gaddam, C. K., Vander Wal, R. L., Chen, X., Yezerets, A., & Kamasamudram, K. Reconciliation of carbon oxidation rates and activation energies based on changing nanostructure. *Carbon*, 2016, *98*, 545-556.

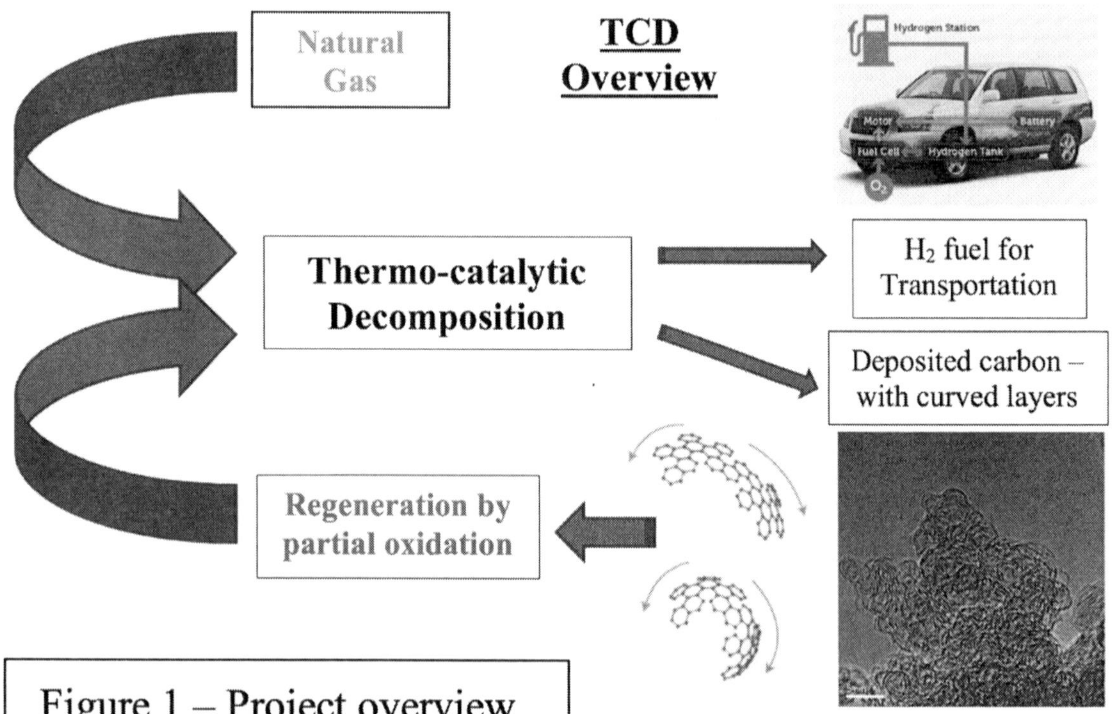

Figure 1 – Project overview.

570b Process Intensified Modular Production of Clean Hydrogen and Carbon Nanomaterials from Natural Gas By Microwave Catalytic Pyrolysis

Changle Jiang, Jianli Hu, Sonit Balyan, Brandon Robinson, Juan A. Lopez-Ruiz, Yuan Jiang, Robert A. Dagle, Sherafghan Iftikhar, Luke Neal, Fanxing Li, Wenyuan Li, Xingbo Liu, George Skoptsov

Microwave catalytic pyrolysis is a feasible pathway for decarbonization natural gas, producing clean hydrogen and valuable carbon nanomaterials without emitting CO_2. More importantly, the process can be portable, modular, and electrification oriented. In this work, we have developed novel and advanced catalysts for attenuating microwave energy, such as multiwalled carbon nanotube supported Ni-based bimetallic nanoparticles (Ni-Pd, Ni-Cu and Ni-Fe), Zr oxides supported Co, exsolved Ni nanoparticles from perovskites matrix (ABTiNiO). Microwave could provide site-specific selective heating with catalyst support temperature being relatively low while the active site temperature being sufficiently high for the methane decomposition reaction. Our data indicated microwave heating helped reduce the activation energy needed for this reaction as well. For example, CNT supported Ni-Cu catalyst can reduce the activation energy of the reaction from 45.5 kJ/mol CH_4 under conventional electrical heating to 24.8 kJ/mol CH_4 under microwave, which was measured with a reaction temperature range from 450 to 600 °C. It was found that the small amount of metal promoter like Cu in the Ni-Cu-CNT catalyst system could modify the carbon nanoproducts from bamboo-shaped CNTs to multiwalled CNTs. In addition, we found that Fe could eventually become a more affordable promoter choice in the CNT support Ni-based catalysts after measurements of activation energies from Ni-Pd, Ni-Cu, and Ni-Fe options. We have also conducted a process simulation and techno-economic analysis of our microwave process that showed the minimum hydrogen selling price was about \$1/kg H_2, which made our process very promising to achieve the goal of Department of Energy to reduce hydrogen production cost to \$1/kg in a decade. In addition, methane plasma pyrolysis product, acetylene, was evaluated with the same CNT supported Ni-based catalyst and showed nearly 100% conversion to hydrogen and carbon nanotubes. Other catalysts such as Zr oxides supported Co and exsolved Ni nanoparticles typically had decent methane conversion rates, but they needed microwave absorbing materials, which were inert to the methane decomposition reactions, as assistance for heating purpose only. Perovskites based Ni catalysts such as SrTiNiO and BaTiNiO had decent conversion of methane and microwave absorbance while the deactivation was faster than those CNTs support Ni-based metal nanoparticles. The work has further successfully separated the metal from CNT product with an acid-wash method. The recovered metal can be used to generated new catalyst. Methane can be converted to acetylene under microwave plasma condition. In a single microwave plasma reactor, acetylene can be converted in the post-plasma zone. The results showed that CNT supported Ni-based and inexpensive alternative Al_2O_3 supported Ni nanoparticles could be a great downstream catalyst for stable conversion of methane derived acetylene to hydrogen and carbon nanotubes. To summarize, the work has developed microwave-suitable catalysts to achieve natural gas catalytic pyrolysis to simultaneously produce valuable carbon nanomaterials and clean hydrogen, which has also been confirmed success in a scaled-up reactor. Our technology possesses a great potential in transforming natural gas to value-added products.

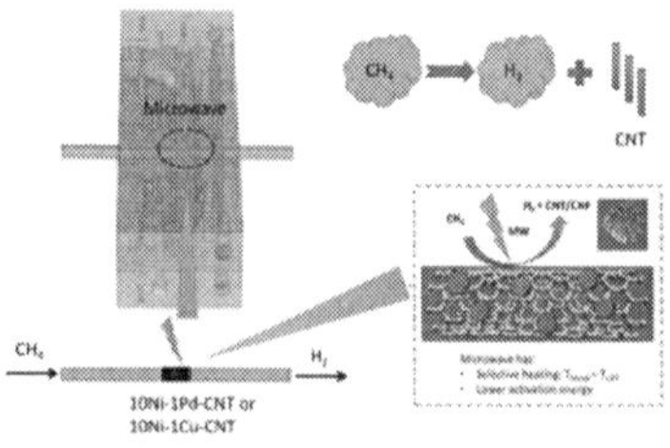
Microwave
CNT
10Ni-1Pd-CNT or
10Ni-1Cu-CNT

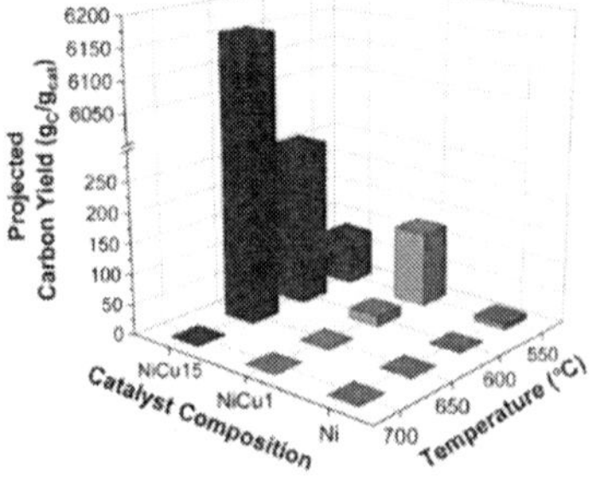
Projected
Carbon Yield (gC/gcat)
6200
6150
6100
6050
250
200
150
100
50
0
NiCu15
NiCu1
Ni
Catalyst Composition
550
600
650
700
Temperature (°C)

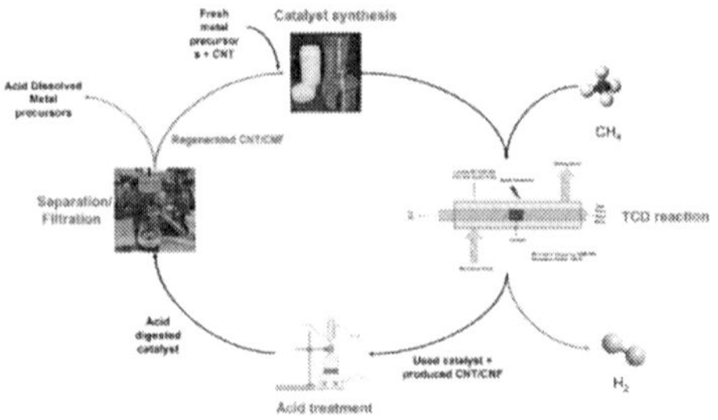
Catalyst synthesis
Separation/
Filtration
TCD reaction
Acid treatment

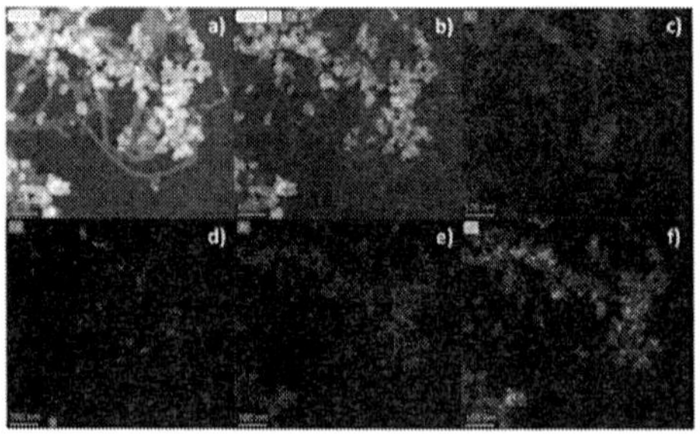
a)
b)
c)
d)
e)
f)

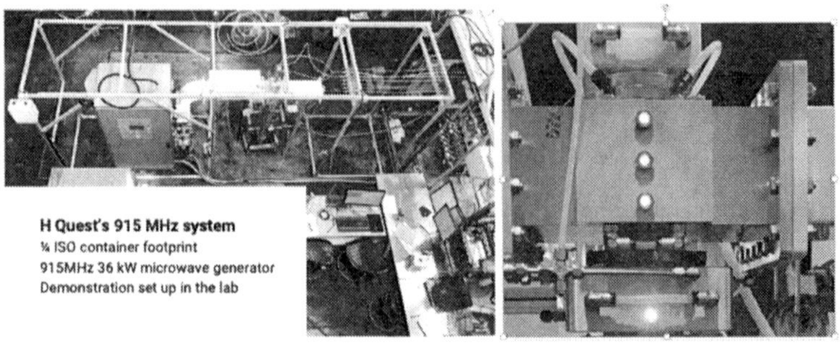
H Quest's 915 MHz system
¼ ISO container footprint
915MHz 36 kW microwave generator
Demonstration set up in the lab

570c Tuning Particle ALD Catalysts for Methane Pyrolysis

Jessica Hauck, Kent J. Warren, Boning Wang, Samantha Harshberger, Mija H. Hubler, Andrew Broerman, Theodore Champ, Christopher Gump, Alan Weimer

Hydrogen is one of the most essential chemical building blocks to our energy infrastructure, with downstream applications in oil refining as well as production of ammonia, the key component in fertilizer. Today, nearly all hydrogen is produced commercially via steam methane reforming. For every kilogram of hydrogen produced, there are roughly 9 kilograms of CO_2 released to the atmosphere.[1] With an estimated global hydrogen production rate of 75 million metric tons per year[1], over 675 million metric tons of carbon dioxide are emitted via this process. A profitable alternative to produce hydrogen while circumventing greenhouse gas (GHG) is needed. A Catalytic-Chemical Vapor Deposition (CCVD) process to produce hydrogen and carbon nanofibers (CNFs) from methane is a financially promising alternative to steam methane reforming.

Methane, the primary component of natural gas, requires substantial process conditions to overcome the strong C – H bonds. Particle Atomic Layer Deposition (ALD) enables the formation of highly dispersed metal nanocatalysts on a given support which reduce the energy required for methane conversion. Moreover, the highly dispersed catalyst, consisting of abundant metals such as iron or nickel, is essential as this process contains a "sacrificial catalyst"; process contains a *sacrificial catalyst*; the catalyst, support, and CNFs combine as one product to improve the strength and durability of concrete. In partnership with both the Hubler group in the department of Civil and Environmental Engineering and the National Ready Mixed Concrete Association (NRMCA), we are investigating the CNF/catalyst product to reduce cracking and improve the durability of high purity concrete (HPC).

In this research, abundant transition metals are deposited onto a support via particle ALD. CCVD and catalyst treatment parameters are investigated to maximize carbon and hydrogen yield. CNFs and hydrogen are then generated via CCVD with the particle ALD catalyst. A preliminary technoeconomic analysis is performed to assess the large-scale viability of this catalytic methane pyrolysis technology.

[1]Bartlett, J., & Krupnick, A. 2020. "Decarbonized Hydrogen in the US Power and Industrial Sectors: Identifying and Incentivizing Opportunities to Lower Emissions" (Report 20-25). *Resources for the Future*.

570d Tailoring the Combustion Synthesis of Fe-Al-O Catalysts for Microwave-Assisted Production of Hydrogen from Fossil Fuels

Zachary Chanoi, Laura A. Martinez Espinoza, Victoria Reyes, Evgeny Shafirovich

Currently, hydrogen is mainly produced from fossil fuels with enormous CO_2 emissions. New technological routes involving electromagnetic energy input are considered promising for clean, low-cost production of hydrogen. In particular, microwave-assisted thermocatalytic dehydrogenation of fossil fuels could produce hydrogen with high yield and selectivity, while generating valuable carbon byproducts and no CO_2 emissions. The success of this technology critically depends on the availability of materials that effectively absorb microwave radiation and catalyze the dehydrogenation reactions. In the present work, the solution combustion synthesis (SCS) of Fe-Al-O compounds was explored to enable the functional design of such materials. The effects of three parameters (fuel, Fe/Al ratio, and heating mode) on the SCS process and products were investigated to reveal process-structure-property relationships important to microwave-assisted catalysis. This study has demonstrated SCS potential to synthesize Fe-Al-O catalysts with a specific surface area up to 250 m^2/g and tunable properties. The combination of the operational parameters that yields the desired phase composition of the products has been determined. The obtained catalysts exhibit excellent absorption of microwave radiation and strong magnetic properties, which could be used for their separation from carbon byproducts.

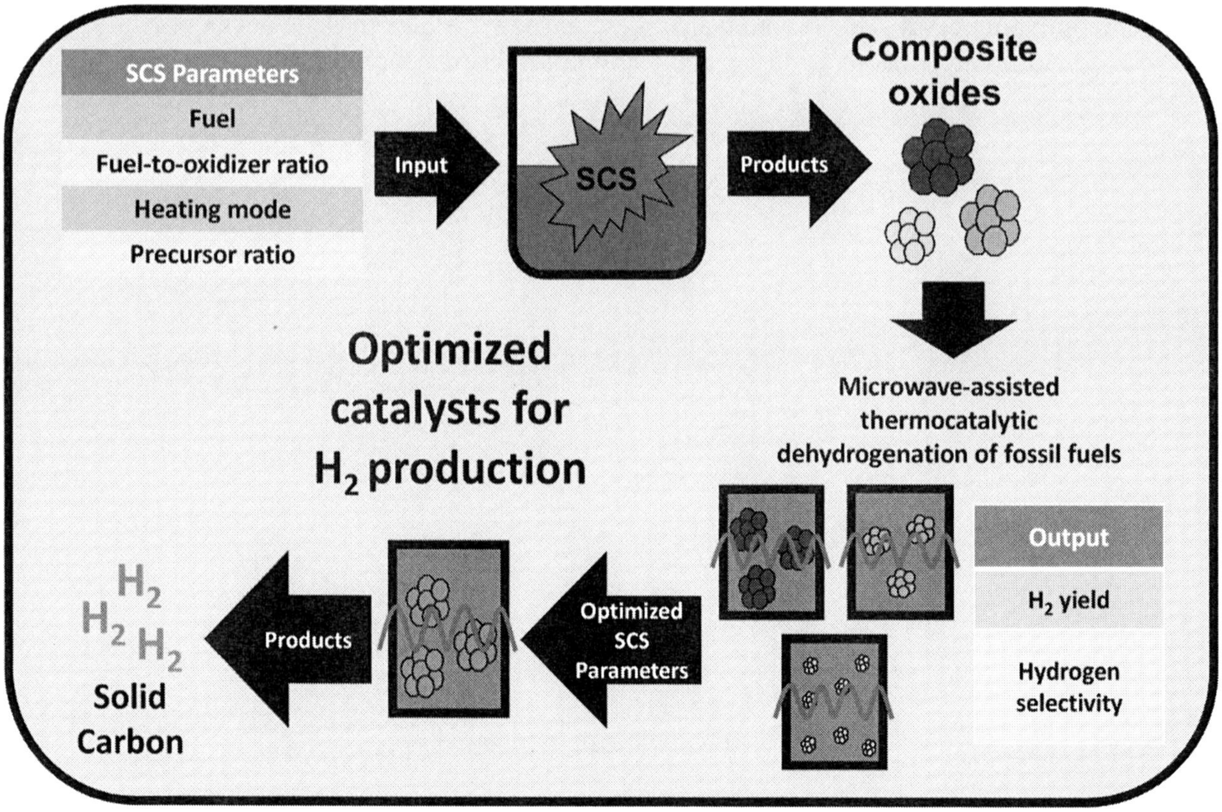

570e Microwave-Assisted Moving-Bed Gasifier Design for Solid Fuel Conversion

Xinwei Bai, Pranjali Muley

Microwave irradiation is an efficient means of heating used for a number of applications in chemical engineering. Driven by the specific characteristics of electromagnetic waves, microwave reactor design leads to an energy-efficient paradigm for energy-intensive processes such as gasification and natural gas conversion. Today, most of the microwave-incorporated reactor designs are either in fixed-bed or batch modes. To address the increasing demand of expanding the capabilities of microwave reactors, this presentation introduces a single-mode flow-type microwave-assisted gasifier system with a co-current moving-bed operation for continuous processing of solid feed such as lignite and plastic wastes. The result from the shakedown using devolatilized lignite shows the system is capable of continuous syngas production for more than five hours. Compared with reactor design using conventional reactors, microwave irradiation improves the overall H_2/CO ratio and produces less carbon dioxide. This study also illustrates the capability of applying this design to plastics upcycling through gasification, and hydrogen-rich syngas production was also observed. The cold gas efficiency was also calculated and compared in this study. In conclusion, this research provides a novel gasifier design which integrates system elements of microwave irradiation and moving-bed operation for continuous solid fuel conversion. These processes can lead to a significant reduction in capital and operating costs. This research also provides new insights into the development of microwave-incorporated reactors and the application of microwave irradiation in modern industry.

570f Modularization Chemical Looping System for Hydrogen Production from Carbonaceous Solid Feedstock Along with CO2 Capture and Additional Heat Recovery

Rushikesh Joshi, Eric Falascino, Andrew Tong, Liang-Shih Fan

Hydrogen is widely regarded as the fuel of the future, but the current supply is primarily derived from fossil fuels. To achieve a more sustainable hydrogen economy, it is essential to develop technologies that can produce hydrogen from both fossil and renewable feedstocks, such as plastics and biomass, while capturing resulting CO_2 emissions.

Chemical looping gasification is a promising approach for the conversion of carbonaceous feedstock into syngas. It typically consists of two reactors, namely the reducer and the combustor. The carbonaceous feedstock enters from the top of the moving bed reducer along with the oxygen carriers. The oxygen carriers donate their lattice oxygen to gasify the carbonaceous species into syngas while both travel downwards co-currently. The generated syngas then leaves from the bottom of the reducer, while the oxygen carriers are sent to the fluidized bed combustor for replenishing the lattice oxygen by air oxidation.

In addition to the two-reactor system, a chemical looping three-reactor system exists for generating hydrogen from carbonaceous feedstocks along with CO_2 capture. In this system, the oxygen carriers react with the carbonaceous species in the reducer to generate a capture-ready stream of CO_2, and the reduced oxygen carriers then travel to the second moving bed reactor, the oxidizer, for steam oxidation. The fluidized bed combustor then fully regenerates the oxygen carriers in the air, closes the heat balance of the system, and conveys the oxygen carriers back to the reducer top. This innovative approach has great potential to reduce greenhouse gas emissions and support the transition to a sustainable energy system.

This study outlines a novel modular chemical looping strategy that utilizes multiple moving bed reactors operating in parallel, connected to a single fluidized bed combustor. The integrated system offers a range of advantages over conventional chemical looping gasification systems, including the option for the co-generation of hydrogen/heat and the elimination of the need for a heat-intensive CO_2 capture unit.

The proposed system can operate in different configurations, providing flexibility to generate syngas or hydrogen based on downstream needs. The system can generate high-purity syngas in the gasification reducer, which can then be converted into hydrogen while also generating a capture-ready stream of CO_2, eliminating the need for an energy-intensive CO_2 capture unit. Alternatively, the system can be integrated into a downstream process that requires syngas as a feedstock, allowing for the co-generation of hydrogen/heat and a separate capture-ready stream of CO_2.

Comparative cases demonstrate that the proposed modular system increases effective thermal efficiency by approximately 17% points compared to conventional chemical looping gasification systems with water gas shift and CO_2 capture units due to the elimination of the CO_2 capture unit and generation of additional heat from the tail gases. Experimental results also demonstrate the feasibility of the proposed process for gasifying biomass and plastics, achieving a high-purity syngas stream for both biomass and plastics gasification on the bench scale moving bed experiments and producing a nearly pure stream of CO_2 for hydrogen generation.

The study also discusses the challenges of dealing with high sulfur content in some fuels, which can be resolved using a syngas cleanup system to recover sulfur. Strategies for the co-injection of plastics and biomass are also presented, showing that the co-injection of polyethylene plastics with biomass can increase the hydrogen yield by approximately 12%.

In addition, the study highlights the advantages of the proposed modular system over the conventional three-reactor Biomass-Direct-Chemical Looping system for hydrogen generation. Specifically, it addresses practical operation and the inherent batch-to-batch variation in biomass. The modular system eliminates the risk of carbon carryover from the reducer to the oxidizer, resulting in high-purity hydrogen, even with varying biomass compositions. Experimental thermogravimetric results on biomass gasification demonstrate the variability in gasification kinetics and how the proposed system can mitigate the risk of carbon carryover.

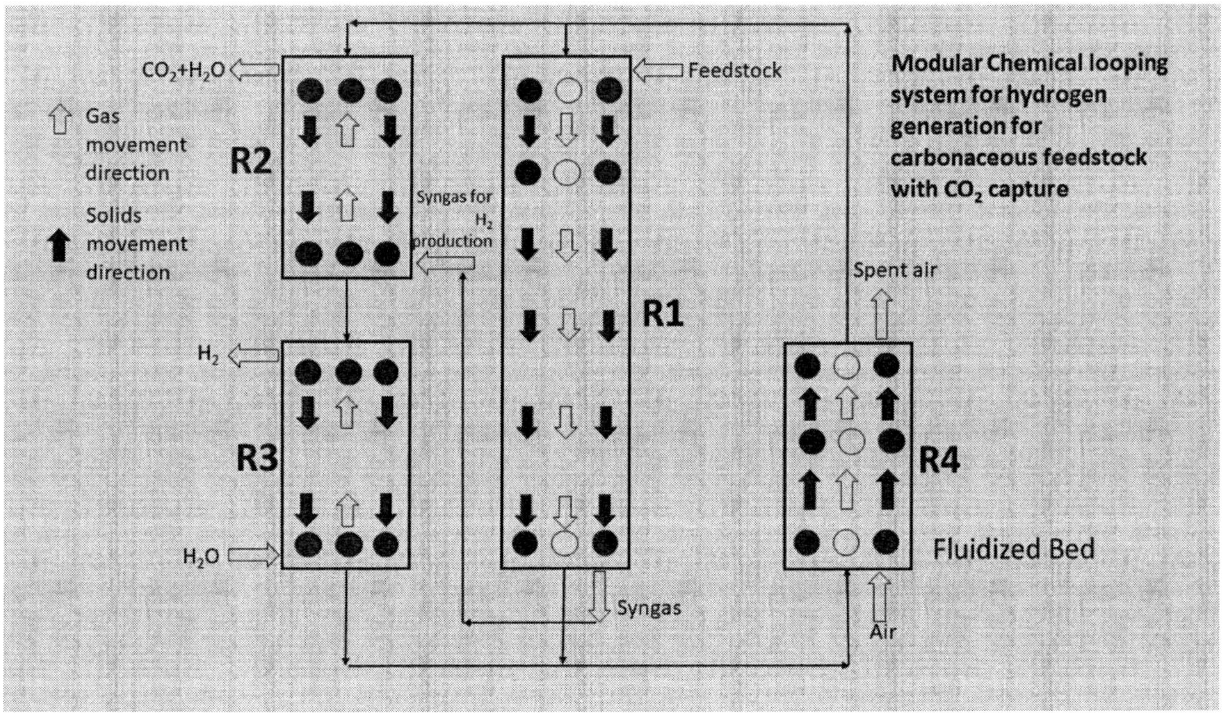

570g Methane Looping Reforming to Syngas: Optimal Process Conditions and Carrier Lifetime Assessment

Alba Storione, Matteo Minelli, Mattia Boscherini, Elena Landi, Francesco Miccio, Ferruccio Doghieri

Globally, synthesis gas (H_2 and CO) and its market has been estimated to expand at a compounded average growth rate (CAGR) of 9.1%, from 2021 to 2031, due to the versatility associated with syngas downstream processing [1].

To date, the mixture of H_2 and CO mainly comes from natural gas, residual oils, petroleum, and coal, as these are the only sources able to cover the world's demand for energy and chemicals.

However, among fossil alternatives, methane is the most investigated reactant, since it is the earth's cleanest burning hydrocarbon, recognized to be a less impacting source for air pollution respect bioenergy or common fuels. Also, natural gas recoverable resources have been estimated to being able to sustain today's production for over 250 years [2].

Currently, the only fully commercialized process based on methane utilization is the steam methane reforming (SMR). However, the critical issues associated with this process, such as the high production costs needed to provide huge amounts of steam at high temperature (650-1000 °C) and pressure (5-40 bar), as well as the severe CO_2 release (about 9 to 12 tons of CO_2 per ton of hydrogen) [3], motivate the study of alternative processes for synthesis gas and hydrogen production.

In this work, a looping process, from methane to syngas, was studied with thermogravimetric analysis characterization (TGA) and experiments in a fixed bed reactor. The process was studied cyclically exploiting the redox properties of a Ce-based oxide oxygen carrier synthetized with a simple forming procedure.

TGA experiments were focused on the identification of the optimal temperatures ranges for methane partial oxidation (900-1000 °C) and carrier regeneration (400-900 °C), while fixed bed testing was performed isothermally (at 900 or 950 °C), with a 10% CH_4 feed stream in N_2 for the stage of reforming and with 3% O_2 in N_2 for the regeneration.

The effect of the process times on carbon deposition, specific syngas yields, and selectivity, was inspected, together with the investigation of best conditions to fully regenerate the carrier, adjust the syngas final ratio, and ensure stable performances.

The obtained results confirmed the possibility to work in fully isothermal operations, with specific yields of syngas per mass of O_2 carrier around 6.8 mmol g^{-1}, preserved even along several cycles, paving thus the path to the development of alternative and effective processes for syngas production.

Reference:

(1) Global Industry. *Syngas and Derivatives Market (Feedstock: Coal, Petroleum, Biomass Waste, and Others; Production Technology: Partial Oxidation, Steam Reforming, Biomass Gasification, and Others; and End-User: Chemicals, Power Generation, Liquid Fuels, and Gaseous Fuels)- Global Industry Analysis, Size, Share, Growth, Trends, and Forecast, 2021-2031.* https://www.transparencymarketresearch.com/syngas-derivatives-market.html.

(2) WEO Special Report: Are We Entering a Golden Age?; 2011. https://www.iea.org/reports/weo-special-report-are-we-entering-a-golden-age.

(3) Lunsford, J. H. Catalytic Conversion of Methane to More Useful Chemicals and Fuels: A Challenge for the 21st Century. Catalysis Today 2000, 63 (2–4), 165–174. https://doi.org/10.1016/S0920-5861(00)00456-9.

Figure: Syngas specific yields and ratio for eight consecutive redox cycles.

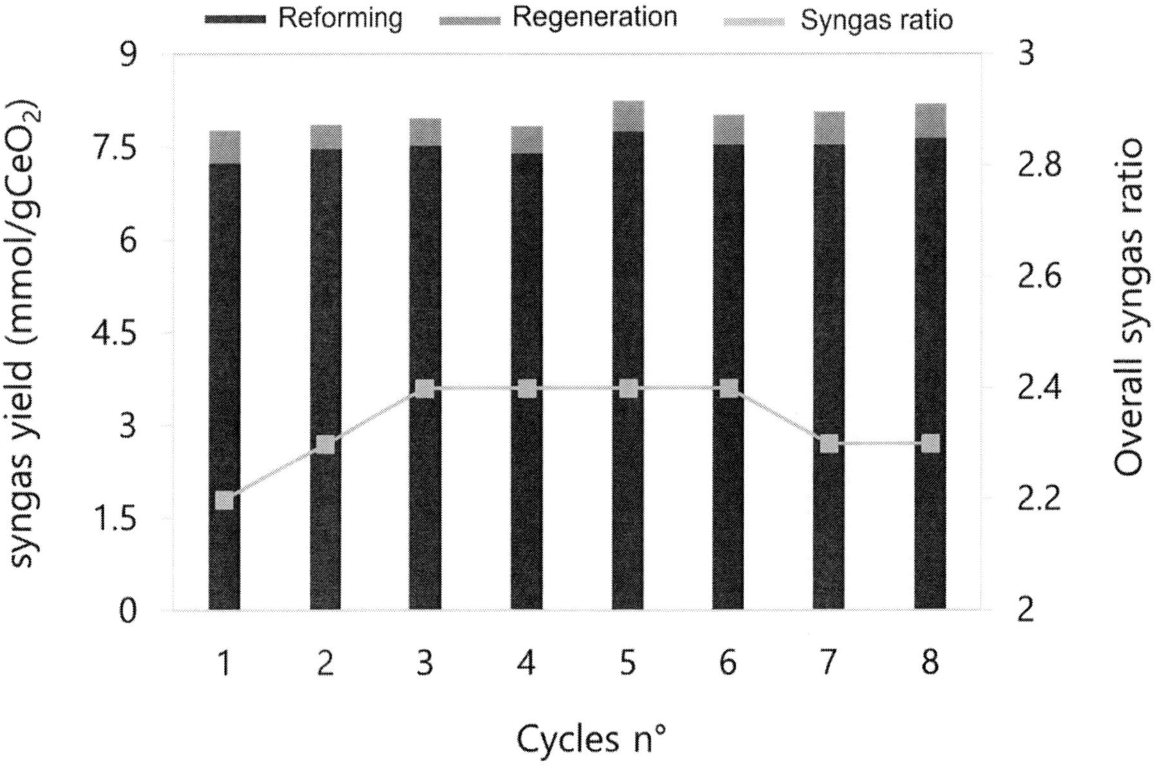

570h Sustainability Analysis of Hydrogen Production Processes for Maritime Transportation.

Andrea Mio, Elena Barbera, Alessandro Massi Pavan, Alberto Bertucco, Maurizio Fermeglia

Hydrogen is a flexible energy carrier and storage medium that can be used for a number of purposes. It might be created through a variety of pathways, each of which would result in a unique type of hydrogen denoted by a colour. In this work, process simulation is used to evaluate capital and maintenance costs, as well as to generate material and energy balances for each process under study. Process simulation outcomes are then used for to estimate three key performance indicators focusing on sustainability pillars: the Energy Return On Energy Invested (EROEI), the Levelized Cost Of Hydrogen (LCOH) and the Life Cycle Assessment (LCA). We have designed the production to feed hydrogen to a ferryboat operating in the North Adriatic sea. We compared several hydrogen generation processes, each denoted by a unique colour code: (i) green hydrogen, produced by electrolysis of water using electricity from renewable sources, (ii) yellow hydrogen, produced by electrolysis using grid electricity, (iii) grey hydrogen, produced from natural gas using steam reforming and (iv) blue hydrogen, like grey one, but with carbon capture and storage (CCS). For green hydrogen it is assumed that the electricity is produced by a floating PV platform located near the utilizers of hydrogen. From the results obtained it is possible to conclude that the most sustainable hydrogen production methods is the so-called green hydrogen, produced by water electrolysis.

526b Preliminary Economic Assessment of Microalgae Via Hydrothermal Liquefaction with Co-Production of Protein Concentrates

Yiling Xu, Yunhua Zhu, Peter Valdez, Lesley J. Snowden-Swan

The present study investigates a process that utilizes microalgae to produce a protein concentrate and generate biofuel through hydrothermal liquefaction (HTL). A preliminary economic analysis was conducted using high-protein algal feedstock to evaluate the benefits and risks of co-producing protein with biofuel, including sustainable aviation fuel (SAF).

Microalgae have been demonstrated to be a promising renewable feedstock for biofuel production. Currently, the high cost of algal feedstock leads to a high production cost for fuels, making it less competitive to other sources. Incorporating a high value co-product is one way to reduce the fuel cost. Microalgae protein is a promising co-product as an alternative protein source (Soto-Sierra et al. 2018). Using microalgae to produce protein-based food/feed augments current supplies for protein products, such as soy and pea. Furthermore, extracting protein from algae prior to HTL offers additional benefits since most of the nitrogen content in the resulting biocrude comes from the protein, which is unwanted in the final jet fuel product. Extracting protein can reduce the nitrogen content in HTL biocrude and thus lowering the hydrogen demand and cost for biocrude upgrading. Therefore, investigating the potential for such protein co-production in the algae-to-fuel process is worthwhile.

In this study, the algae feedstock is extracted to produce protein concentrate and then the protein-extracted algal residue is converted into biocrude through the HTL process. Finally, the biocrude is hydrotreated, hydrocracked, distilled and denitrogenated to produce naphtha, diesel, and SAF range fuels. For the baseline case, the algae feedstock is assumed to be from a 3400-acre open-pond farm, producing algae at a cost of 1,000 \$/ton on an ash-free dry weight (AFDW) basis. With protein extraction and HTL conversion, the modeled fuel yield is approximately 49 gasoline gallon equivalents (GGE) per ton algae on AFDW basis. The produced fuel products were composed of 60% SAF and a balance of diesel and naphtha fractions. Results show that in the baseline case, the minimal fuel selling price (MFSP) can be lowered below \$4/GGE if the selling price of protein concentrate is above \$1.4/lb. A sensitivity analysis shows that algae feedstock prices and plant scale have the most influence on the MFSP, which can vary from \$0.5/GGE to \$25/GGE. Alternative case studies, including algae conversion to fuel only (without protein extraction) and algae/wood blended feedstock to fuel only, are also evaluated and compared to the baseline case.

Reference:

Soto-Sierra, L., P. Stoykova, and Z.L. Nikolov, Extraction and fractionation of microalgae-based protein products. *Algal Research*, 36, 2018; 175-192

526c Sustainable Biolubricant Production from Lignin and Waste Cooking Oil

Kyoko Hirayama, Elena Barbera, Randy Maglinao, Alberto Bertucco, Sandeep Kumar

Lignin accounts for 15-30% of lignocellulosic biomass weight and 40% of the biomass heat energy. Today, lignin from lignocellulosic biorefineries is underutilized because of the absence of a high-value product from wet unhydrolyzed solids (UHS), a waste stream that has very little value. Finding ways to valorize this byproduct is crucial to enhance the economic viability and sustainability of biofuels production.

In this work, a process is developed to transform lignin, or UHS, into a sustainable biolubricant. The development of sustainable and environmentally friendly lubricants has in fact become an important challenge, and their market has increased significantly in recent years. Fatty acids and their corresponding methyl esters (or FAMEs) are good candidate as biolubricants, however those derived from plant oils do not have desirable oxidative stability due to the presence of C=C bonds that make them susceptible to autooxidation. The stability can be improved by chemically modifying the fatty acid molecule, utilizing the C=C bond's reactivity to create linkages or new molecules with a branched structure.

The proposed process consists of three steps: (1) Hydrothermal liquefaction (HTL) of lignin (UHS), to depolymerize it into monomeric phenolic compounds, collected into a bio-oil; (2) Hydrodeoxygenation (HDO) of the bio-oil to replace the hydroxyl group of phenolic compounds with hydrogen, and produce aromatic hydrocarbons; (3) alkylation of unsaturated FAMEs (in particular from transesterification of yellow grease) with the aromatic hydrocarbons obtained from lignin degradation.

HTL was conducted at temperatures between 280-320°C, and reaction times between 1 - 15 min using an extremely efficient induction heating system that allowed reaching the reaction temperature within 2-3 min. The optimal operating conditions that allowed to maximize the bio-oil yield were found to be 320 °C at 1 min residence time, and the oil characterization revealed that in these conditions the selectivity of the phenol was highest at 52% of relative peak area with GCMS analysis. Furthermore, the fast HTL was found to increase the bio-oil yield and phenol content in it significantly compared to those conventional HTL which take pre-heating time of 60 minutes.

Hydrodeoxygenation was performed at 250°C and 600 psi of hydrogen gas. It was hypothesized that the polarity of the solvent affects the aromatic selectivity of the hydrodeoxygenation process. An alkane (i.e., *n*-hexadecane) and water were selected as nonpolar and polar solvent, respectively. Based on the experimental results using *p*-cresol as a model compound, aromatic selectivity was significantly higher when hexadecane was used as a solvent. Thus, hexadecane was used in the HDO of biooil obtained from HTL step. Alkylation of the aromatic hydrocarbons with unsaturated FAMEs was conducted using an acid-treated montmorillonite catalyst at 210°C for 4 hours under argon gas.

Finally, a techno-economic analysis was conducted based on the experimental data collected, to verify the scalability and economic viability of the proposed process when integrated in an already existing lignocellulosic biorefinery.

526d A Process for Co-Conversion of Bio-Oil in Delayed Coker Unit As Viable Option for Achieving Net Zero Goals

Satyen Kumar Das, Shivam Ashok Dixit, Pradeep P.R., Sarvesh Kumar, Madhusudan Sau, S. S. V. Ramakumar

Petroleum refiners are in the process of identifying and planning for adoption of new technologies for transitioning to 'Net Zero'. Increasing the use of alternate carbon-neutral feedstocks like biomass etc. in the existing petroleum refineries is being projected as one of the potential paths to achieve the same. International Energy Agency estimates that liquid biofuel consumption will increase from current levels of 1.6 million barrels of oil equivalent (mboe/d) in 2020 to 6 mboe/d in 2030 in the 'Net Zero' Scenario. Though several bio-refining technologies are under developed or are being developed, it has been observed that these processes require separate process plants, such as set up for biomass processing and purification of products. Owing to the higher energy density of bio-oil, biomass pyrolysis process is deemed as a preferable route for overcoming the issues associated with biomass processing. In this backdrop, it is desirable to have technologies that could enable the petroleum refiners to process biomass derived feedstocks within the existing petroleum refining infrastructure to enable faster adoption of sustainable energy technologies.

In view of the above, a process has been developed for co-conversion of biomass pyrolysis oil in refinery Delayed Coker unit (DCU) to produce sustainable low carbon fuels. Biomass pyrolysis oil is characterized by its unique features like high Conradson Carbon Residue content akin to heavy petroleum residues, processed in DCU. Bio-oil derived from fast pyrolysis of lignocellulosic biomass is a complex mixture of oxygenated compounds, which poses challenge for refinery processing. To evaluate the impact of bio-oil in varying proportion in conventional DCU feedstock, experimental study was undertaken using lignocellulosic biomass such as Jatropha, Karanjia, Rice straw & Wood based biomass pyrolysis oil in Batch Coker Unit and validated in 1 bbl/day DCU pilot plant unit. The experimental results indicate that inclusion of bio-oil in feed to DCU has several benefits including potential to reduce the coke yield & improve the yield of distillates. It was also observed that Bio-oil dosing in DCU feedstock depends on the quality of Bio-oil derived from biomass sources.

This paper describes development of a process for co-conversion of bio-oil in Delayed Coker unit for sustainable low carbon fuel production from waste biomass pyrolysis oil aiming towards **'Path to Net Zero'.**

Evolution and Development of Fluidized Catalytic Dehydrogenation (FCDh) Technology

Matt Pretz

For decades, the world has been in search of an economic Circulating Fluid Bed (CFB) Technology for the production of olefins through Fluidized Catalytic Dehydrogenation (FCDh). Dow's UNIFINITY(TM) FCDh Technology is the culmination of the world's search and is now available for license by Univation Technologies, LLC, a wholly owned subsidiary of Dow. This paper will describe FCDh's evolution from the original bubbling bed reactor concepts introduced in Russia by Yarsintez and later improved upon by Snamprogetti for the production of C4's to the modern day Fluidized Catalytic Dehydrogenation (FCDh) technology commercialized by Dow for the production of propylene. The reasons for the migration from bubbling bed reactor to modern day fast fluidized/turbulent bed reactors with short contact time separation will be discussed. In addition, some of the conceptual tools Dow has used in scale-up of the technology from laboratory scale to full scale production will be described. The presentation will conclude by updating the technical community on the status of the Dow FCDh plant for the production of propylene in Plaquemine, Louisiana.

483b Ethylene Oxide Catalyst and Process Innovations: A Successful Journey Towards Asustainable Industrial Process

Madan M. Bhasin

Ethylene oxide (and ethylene glycol) are very large volume, commodity chemicals and building blocks for a large number of other derived chemicals-with annual production exceeding 50 Billion pounds per year. These commodity chemicals have been growing globally at the rate of 5-6%/year for the last 3-4 decades. Improving catalyst selectivity and hence reducing carbon dioxide emissions has been very intense competitive activity amongst the major competitors; Union Carbide/Dow Chemical, Shell/Criterion and Scientific Design for the last 50 years. The key catalyst inventions leading to innovations of successively improved generations of ethylene oxide catalysts from about 70% to 90%, over the last three/four decades, will be presented. In addition, the process evolutions from the old multi-reactors in series for Air based and then the multiple parallel reactor-based oxygen-based processes leading to the current designs of a single reactor to produce 500-1000 MM lbs/year in a single plant, will be presented. In addition, these innovations have resulted in a much simpler process design having lower capital & operating costs along with
process simplicity and enhanced process safety. Most importantly, from sustainability standpoint, Carbon dioxide emissions have been reduced enormously, from about 30% to 10%, even though the worldwide capacity has nearly doubled in the last 15 years. Importantly, even more reductions in carbon dioxide emissions and improvement in selectivity are possible in the coming years.

483c Imagine This a Tribute to Bipin Vora on His 80th Birthday

James Johnson

In his decades-long career with UOP, Bipin has been one of the most inventive and entrepreneurial Chemical Engineers in the company's history. He didn't just come up with great ideas, he advocated for getting them developed and drove them to be commercially proven. His efforts have had seminal and long lasting impact across several industries: clean fuels, bio-degradable detergents (Detal™, DeFine™), , on-purpose production of propylene (Oleflex™) and methanol to olefins (Advanced MTO™) – to name a few. This presentation will highlight some of his key technical accomplishments, the worldwide recognition he has achieved, and the impact he has had on the lives of young future scientists and engineers around the world.

483d Strategies for Improved Thermal Stability and Facile Regeneration of Alkane Dehydrogenation Catalysts

Abhaya Datye, Ryan Alcala, David P. Dean

One of Bipin Vora's most recognized contributions [1] is the development of a light paraffin dehydrogenation process, named Oleflex^{TM}. This process was first commercialized for propylene production in 1990. UOP Honeywell has licensed 80 units for propane and isobutane dehydrogenation for the production of propylene and isobutene and accounts for 75 % of the worldwide propane catalytic dehydrogenation capacity added in the last 10 yrs. Bipin Vora's contributions to Oleflex are documented in 25 US patents and numerous publications. He made key inventions and conducted pioneering research that is the cornerstone of the current and future generations of UOP Oleflex technology.

Activating a C-H bond and converting an alkane selectively to alkene is a challenging reaction which is carried out at high temperatures due to thermodynamic limitations. During the reaction, the olefins that are formed can react further to undesired products involving hydrogenolysis and also result in coke deposits that can lead to catalyst deactivation. The high temperatures also cause sintering of the active phase and loss of activity. In industrial practice, the catalysts are regenerated using a controlled amount of oxygen and in the presence of halides. At the NSF Engineering Research Center CISTAR [2], we are exploring process technologies that could be applied for the processing of natural gas liquids closer to the well head. In the context of distributed processing, alternate approaches for regeneration need to be explored. For instance, the use of air instead of diluted oxygen for regeneration and avoiding the need for added halides. Prolonging the lifetime of these catalysts would also be beneficial for distributed processing. Here, we demonstrate that the choice of support and the composition of the active phase may allow improved thermal durability and facile regeneration of alkane dehydrogenation catalysts [3].

[1] Vora, B.V. *Development of dehydrogenation catalyst and processes*; Topics in Catalysis, 2012, 55 pp1297-1308.

[2] NSF Engineering Research Center, Center for Innovative and Strategic Transformation of Alkane Resources (CISTAR), www.cistar.us

[2] Alcala, R., Dean, D.P., Chavan, I., Chang, C.-W., Burnside, B., Pham, H.N., Peterson, E., Miller, J.T., and Datye, A.K., Strategies for regeneration of Pt-alloy catalysts supported on silica for propane dehydrogenation. Applied Catalysis A: General, 2023. 658: p. 119157.

483e Methanol-to-Olefins (MTO). Bipin Vora's Key Role in the Development of the UOP/Hydro MTO Process Technology

Steinar Kvisle, James Johnson

My association with Bipin Vora started when Norsk Hydro of Norway and UOP formed a team for the joint development of the Methanol to Olefins technology in 1991. UOP had invented a new group of materials, silica alumina phosphates (SAPO), and had shown that in particular the SAPO-34 material was very selective for the methanol to olefins reaction. On the UOP side, it was then the task of Bipin Vora to secure R&D funding from the management of UOP. Bipin Vora was successful for having a budget for the project based on his global outlook for the future availability of low-cost methanol. Norsk Hydro developed a similar view in the late 1980'ies on future availability of low-cost methanol, especially based on stranded Natural Gas, and saw Methanol-To-Olefins as a possible new route since this would open up for a new feedstock for the production of polymers. Through introductory laboratory work we con-firmed the potential of SAPO-34 as a catalyst for the methanol to olefins reaction. Therefore, since Norsk Hydro shared the same vision for MTO as developed by Bipin Vora in UOP, Norsk Hydro joined forces with UOP in 1991 for the development of the technology. Hydro's most important contribution to the development has been the construction and operation of an MTO Demo Plant in Norway. In 2008, the internationally leading chemical company INEOS acquired Norsk Hydro's petrochemical business and the MTO business was transferred to INEOS (Inovyn). Bipin Vora was a continuous and very constructive force behind the UOP/ Norsk Hydro MTO team till he retired at the end of 2006. This presentation will highlight the key role Bipin Vora played in the joint development of the MTO-technology, from the early start till the technology was announced available for licensing in 1995 and thereafter the successful commercialization of MTO Process.

AUTHOR INDEX

Abatzoglou, Nicolas 63
Achouri, Esma Ines 61, 63
Adinolfi, Michele 118
Aduev, Boris 90
Ahring, Birgitte 6
Akhter, Sayma 72
Alcala, Ryan 148
Al-Douri, Ahmad 31
Aljeshi, Majed 124
Allegro, Alexis 9
Almallahi, Rawan 106
Al-Rawashdeh, Mamoun 95, 113
Al-Shehri, Dhafer 87
Alyousef, Zuhair 124
Amar, Vinod 71
Antoniv, Marta 18
Arges, Christopher 123
Asama, Micheal 117
Ashour, Faisal 95
Aston, John E. 71
Ayala, Rosely 88
Baffoe, Janet 60, 62
Bai, Xinwei 136
Baled, Hseen O. 15
Balyan, Sonit 132
Barbera, Elena 118, 141, 143
Bell, Elijah 16
Bello, Kolawole 54, 102
Belokurov, Gennadiy 90
Bertucco, Alberto 118, 141, 143
Bharanidharan, Aaditya Hari 108
Bhasin, Madan M. 146
Bhattacharjee, Gaurav 51
Bhosale, Rahul 72
Biswas, Sayandeep 26
Bliznakov, Stoyan 11
Boscherini, Mattia 139
Brandenbusch, Christoph 98
Broerman, Andrew 134
Brooker, Paul 122
Broussard, Dylan 38
Brown, Sean 110
Burgess, Ward A. 15
Buttner, William 31
Caiola, Ashley 110
Castaldi, Marco J. 39
Ceja, Marco 88
Chakraborty, Alisa 123
Champ, Theodore 134
Chang, Sehoon 17
Chanoi, Zachary 135
Charalambous, Margarita 35
Cheekatamarla, Praveen 29
Chen, Cong 100
Chen, Hsiang-Sheng 64, 111
Chen, Xueli 10
Cheng, Yi 66
Creadore, Lauren 39
Cunha, Luciane 102
Dagle, Robert A. 132
Dai, Sheng 116
Daraboina, Nagu 16
Das, Satyen Kumar 144
Datye, Abhaya 148
Dean, David P. 148
Deshavath, Narendra Naik 2
Dixit, Shivam Ashok 144
Doghieri, Ferruccio 139
Drummond, Samuel 9
Du, Ruiqi 66
Espinoza, Laura A. Martinez 135
Falascino, Eric 137
Fan, Liang-Shih 137
Fan, Xin 88
Fazio, James 15
Febrianto, Wiwin 53
Fenton, James 122
Fermeglia, Maurizio 118, 141
Fickbohm, Zoe 70
Gamwo, Isaac 15
Gbadamosi, Afeez 23, 87
Giduthuri, Anthony 6
Giles II, Jefferson 96
Gomez, Joshua 47
Green, William 26
Groth, Katrina 31
Guillén-Gosálbez, Gonzalo 35
Gump, Christopher 134
Gunukula, Sampath 30
Guo, Mond F. 6
Gupta, Ram 71
Habicht, Jonas 98
Han, Donggu 48
Haribal, Vasudev 107
Harisankar, S 114
Harold, Michael 104

Harshberger, Samantha 134
Hartman, Ryan 53
Hartmann, Kevin 31
Hassiba, Raid J. 113
Hassini, Hanaa 61
Hauck, Jessica 134
Hawa, Hani 59
Henckel, Danielle 49
Herrera, Megan 116
Hirayama, Kyoko 143
Hoekstra, Hsiang Yee 1
Hong, John 59
Hsiao, Li-Yin 55
Hu, Jianli 110, 132
Hu, Leiming 49
Huang, He 84
Hubler, Mija H. 134
Hunt, Savannah G. 28
Hussain, Syed Muhammad Shakil 87
Hussain, Syed 23
Huynh, Khang 70, 71
Hwang, Hyun-Tae 28
Iftikhar, Sherafghan 107, 132
Igenegbai, Valentina Omoze 106
Ikbarieh, Abdallah 116
Ishida, Serina 88
Islam, Md Tahmid 116
Ismagilov, Zinfer 90
Italiani, Eduardo 68
Iwai, Hiroshi 44
Jeffryes, Clayton 20
Jeong, Sang Mun 89
Jiang, Changle 132
Jiang, Yuan 132
Jin, Wencheng 116
Jin, Zhehui 100
Johnson, James 147, 149
Jones, Rob 26
Joshi, Rushikesh 137
Junaedi, Christian 59
Kamal, Muhammad Shahzad 23, 87
Karakache, Sabrina Bahia 63
Karunarathne, Sampath 30, 34
Kazantzis, Nikolaos 68
Khanal, Aaditya 117
Kilaz, Gozdem 10
Kim, Byung Uk 28
Kim, Geo Jong 28
Kim, Sang Bum 28
Kim, Tae Jin 11
Kishimoto, Masashi 44
Kline, Matthew 34
Klinger, Jordan 116
Koh, Carolyn 53
Kraft, Yaroslav 90
Kumar, Gopalakrishnan 72
Kumar, Sandeep 71, 143
Kumar, Sarvesh 144
Kvisle, Steinar 149
Kwak, Hyung 18
Landi, Elena 139
Lauterbach, Jochen 9
Leclerc, Corey 60, 62
Lee, Ha Eun 89
Lee, See Hoon 89
Lee, Ung 48
Li, Anni 84
Li, Fanxing 107, 132
Li, Jie 21
Li, Weizhong 100
Li, Wenyuan 132
Li, Xiujuan 84
Ling, Jester Lih Jie 89
Linga, Praveen 51
Linic, Suljo 106
Linke, Patrick 95, 113
Liu, Guoxiang 54, 102
Liu, Junchen 107
Liu, Junli 10, 97
Liu, Xingbo 132
Liu, Zhongzhe 88
Lopez-Ruiz, Juan A. 132
Lozano, Robert 88
Maddipudi, Bharathkiran 70, 71
Magee, Joseph 99
Maglinao, Randy 143
Mahalec, Vladimir 14
Makinde, Oluwanisola 119
Martinez, Louis Edwards Caceres 10
Masnadi, Mohammad S. 108
McGaughy, Kyle 108
Miccio, Francesco 139
Millspaugh, Allison 28
Minelli, Matteo 139
Mio, Andrea 118, 141
Mohaghegh, Shahab D. 54, 102
Molina, Elena Robles 10
Monte, Gia 108
Moreno-Sader, Kariana 26
Mosier, Nathan S. 10
Motagamwala, Ali Hussain 106
Muley, Pranjali 136
Muradov, Nazim 64, 111
Nakamura, Riki 44

Neal, Luke ... 107, 132
Nelubina, Natalia ... 90
Neyerlin, K. C. ... 49
Nkiawete, Mpila ... 129
Nurmukhametov, Denis ... 90
Obijiaku, Jude C ... 120
Ocampo, Kaylen ... 108
Oh, Seung Seok ... 89
Oke, Mayokun ... 7
Olotu, Kristinoba ... 5
Onsree, Thossaporn ... 9
Onyelucheya, Chioma M ... 120
Onyelucheya, Okechukwu ... 120
Ouyang, Qian ... 52
Page, Jeffrey ... 11
Pan, Qiong ... 21
Pandey, Jyoti Shanker ... 52
Park, Chan Seung ... 88
Park, Han Saem ... 89
Park, Hyun Jun ... 89
Patil, Shirish ... 23, 87
Pavan, Alessandro Massi ... 141
Pawar, Prasad ... 20
Pendse, Hemant P. ... 30
Pham, Xuan ... 60, 62
Poitzsch, Martin ... 18
Pophali, Amol ... 11
Pradeep, P. R. ... 144
Pretz, Matt ... 145
Qian, Shuairen ... 66
Ramakumar, S. S. V. ... 144
Ramani, Vijay ... 64, 128
Ramasamy, Karthikeyan ... 6
Ratcliff, Jonathan ... 104
Razzak, Shaikh ... 43
Resende, Fernando ... 5, 7, 117, 119
Reyes, Victoria ... 135
Reza, Toufiq ... 116
Riazi, M. R. ... 14, 50
Robinson, Brandon ... 110, 132
Rodriguez, Jose ... 103
Romanov, Vyacheslav ... 54, 102
Roy, Partho ... 88
Roychoudhury, Subir ... 59
Sadowski, Gabriele ... 98
Saha, Nepu ... 116
Saha, Prantik ... 49
Saha, Rahul ... 24
Salami, Oluwaseun ... 34
Santos, Gabriel ... 16
Sanyal, Udishnu ... 6
Sarica, Cem ... 16
Sau, Madhusudan ... 144
Saur, Genevieve ... 31
Schoch, Alexander ... 35
Schulte, Lisa ... 67
Schwab, Steve ... 56
Setzler, Brian P. ... 121
Shafirovich, Evgeny ... 135
Shahabuddin, Muntasir ... 68
Sharma, Kritika ... 64, 128
Sharma, Mrityunjay ... 53
Shell, Katelyn ... 71
Shende, Anuradha ... 71
Shende, Rajesh ... 70, 71
Sheng, Yijie ... 84
Singh, Vijay ... 2
Skoptsov, George ... 132
Snowden-Swan, Lesley J. ... 142
Srinivas, Girish ... 56
Stathatou, Patritsia M. ... 13
Storione, Alba ... 139
Subramaniam, Ramalingam ... 38
Sun, Jingyue ... 100
Tao, Bernard ... 97
Teixeira, Andrew R ... 68
Thakkar, Anuj ... 71
Thies, Mark C. ... 96
Thomas, Gawain ... 17, 18
Timko, Michael T. ... 68
Tong, Andrew ... 107, 137
Twizerimana, Aime Laurent ... 108
Valdez, Peter ... 142
Valla, Julia A. ... 11
Van Cleve, Tim ... 49
Vander Wal, Randall ... 129
Veser, Goetz ... 108
Vijayan, Varshith ... 38
Vikara, Derek M. ... 54, 102
Vilekar, Saurabh A. ... 59
Vinu, R. ... 114
Vogt-Lowell, Kyle ... 107
Von Solms, Nicolas ... 52
Wang, Boning ... 134
Wang, Minghui ... 84
Wang, Wei ... 17
Wang, Xiaoyan ... 110
Wang, Yifan ... 122
Wang, Yuxin ... 110
Warren, Kent J. ... 134
Weaver, Helena Hagelin ... 55
Weimer, Alan ... 134
Weiss, Catherine M. ... 121
Wheeler, Clayton ... 30, 34

Wild, Katherine...67
Wortman, James..106
Wright, Mark Mba...67
Wu, Jiangtao..99
Wu, Xiongjun...54
Xu, Yiling..142
Yan, Yushan..121
Yancey, Neal...116
Yang, Yuan...53
Yelvington, Paul...64, 111
Yik, Edwin...64, 111
Zhang, Kaizheng..66
Zhang, Ye...51
Zhao, Siqi...66
Zhou, Wendy..111
Zhu, S. Sherry..18
Zhu, Yunhua..142

AIChE
120 Wall Street, FL 23
New York, NY 10005-4020

ISBN 978-1-7138-9306-6
9 781713 893066